The Psychology of Music and Autism

Pamela Heaton

The Psychology of Music and Autism

Hearing, Feeling, Thinking, Doing

Pamela Heaton
Department of Psychology
Goldsmiths University of London
London, UK

ISBN 978-3-031-70402-4   ISBN 978-3-031-70403-1   (eBook)
https://doi.org/10.1007/978-3-031-70403-1

© The Editor(s) (if applicable) and The Author(s), under exclusive licence to Springer Nature Switzerland AG, part of Springer Nature 2024

This work is subject to copyright. All rights are solely and exclusively licensed by the Publisher, whether the whole or part of the material is concerned, specifically the rights of translation, reprinting, reuse of illustrations, recitation, broadcasting, reproduction on microfilms or in any other physical way, and transmission or information storage and retrieval, electronic adaptation, computer software, or by similar or dissimilar methodology now known or hereafter developed.
The use of general descriptive names, registered names, trademarks, service marks, etc. in this publication does not imply, even in the absence of a specific statement, that such names are exempt from the relevant protective laws and regulations and therefore free for general use.
The publisher, the authors and the editors are safe to assume that the advice and information in this book are believed to be true and accurate at the date of publication. Neither the publisher nor the authors or the editors give a warranty, expressed or implied, with respect to the material contained herein or for any errors or omissions that may have been made. The publisher remains neutral with regard to jurisdictional claims in published maps and institutional affiliations.

This Palgrave Macmillan imprint is published by the registered company Springer Nature Switzerland AG.
The registered company address is: Gewerbestrasse 11, 6330 Cham, Switzerland

If disposing of this product, please recycle the paper.

Preface

Writing a book on music and autism has been something I planned to do for many years, though my ideas about the form this might take took a long time to evolve. My earliest childhood interest was in music. I began to study the piano when I was five years old and sang in choirs and as a soloist throughout my childhood and early adolescence. As a young adult I began to study music seriously, primarily focusing on developing my skills as a classical singer. When I later changed direction and completed undergraduate and postgraduate degrees in psychology my earlier musical experiences influenced both my research direction and my beliefs about what this research might achieve.

My interest in studying music in autism emerged after I came to know the late Beate Hermelin, who was then working with Linda Pring at Goldsmiths University where I was a student. I later joined Beate and Linda as a research assistant on their project investigating the savant syndrome. As a post-graduate psychology student with a background in the arts, this was a life-changing opportunity. This post enabled me to work with savant artists and musicians and the families and professionals who supported and worked with them. In their work Beate and Linda acknowledged the multidisciplinary nature of the questions they were addressing and this involved working with people with additional expertise. For example, in their study of the savant artist Richard Wawro they involved the psychiatrist Sula Wolff, who had worked with Richard as a child, and the artist Michael Buhler, who provided invaluable insights on the study design and our interpretation of the ideas Richard expressed in his artwork. During this period I was also fortunate in being allowed to work in a wonderful West London School that educated autistic children and adolescents. I will forever be grateful for the opportunities that Beate, Linda and the staff at this school afforded me. But most important for my understanding of musicality in autism was what I was able to learn from the children I met at the school.

At the time I carried out my early research into autism and music, cognitive psychology was the dominant paradigm. My work focused on exploring the perceptual and cognitive processes then believed to be the 'building blocks' of musical strengths in autism. However, I did manage to carry out a small-scale study of the children's perception of emotions in music, and in retrospect I believe this produced the most interesting finding from my early research.

After my research studies were published I began to correspond with a wide group of autistic people, family members and professionals who had read these papers and shared my interests. Some of these correspondents strongly influenced my understanding of the importance of music for autistic people. Most notable was Reubs Walsh, now an established researcher, whose insights into the importance of music for interpreting and conveying emotions resulted in a joint paper written by Reubs, the psychologist (and coincidentally, my husband) Rory Allen and the philosopher Nick Zangwill. From my later correspondence and person-to-person meetings with the wonderful autism+music experts who feature in this book, Andrew, Elisabeth, James, Sarah, Stephanie and Will, I was able to learn about the complexity and variability in the trajectories of musical development that result in expertise within this domain. Their perceptions and beliefs also influenced the way I approached the literature on changing perceptions of autism and on music and musicality in this book. I have used the term autism+music experts to refer to them and feel I should explain why I have done this. During my research I encountered the important work of Monica Botha and other autistic scholars writing on autism and personal identity. When I discussed this with some of my correspondents they described how their musicality as well as their autism were crucial aspects of their personal identity. Hence my use of the term autism+music experts.

Traditionally, within psychology, groups of autistic individuals have been studied as 'subjects' or 'participants' and theories have been created within academic communities that have not included autistic people. However, the limitations of approaches that fail to acknowledge the lived experiences of these so-called subjects or participants have become strikingly clear in the case of autism. The greatest advances in understanding musicality in autism have resulted from studies reporting on first-person accounts of musical engagement, and I describe some of this work in Chap. 4. My hope is that this book will extend these findings to gain a better understanding of this unique aspect of human experience in autistic individuals with exceptional musical talent.

London, UK   Pamela Heaton

Contents

1 Introduction to Autism and Music   1
Autism and Autism Spectrum Disorder   2
Music, Musicality and Musical Expertise   4
Cognitive Models of Autism   6
Cognitive Models of Talent in Autism   8
Rethinking Thinking About Musicality in Autism   10
References   12

2 Conceptualising Musicality as a Complex Human Trait   19
Musicality and the Biomusicology Framework   19
Causation or Mechanism   21
Ontogeny or Development   22
Evolution   25
Function or Adaptation   29
Summary   30
References   30

3 Evolving Concepts of Autism   37
Brief History of Autism   37
Changes in the Diagnostic Classification of Autism   39
Approaches to Understanding Autism   43

	Research Into Early Development in Autism	44
	Changing Perspectives	47
	References	49

4 Music and Autism: Time for a Reappraisal? 59
Contextualising Musicality in Autism 59
Changing Concepts of Autism 62
The Uses of Music in Everyday Life in Autism 64
References 69

5 First-Person Accounts of Musical Talent in Autistic Adults 75
Musical Journeys 76
References 83

6 Musical Journeys 1: Monotropism, Flow and Musicality 85
Monotropism 86
Flow 86
Flow and Music 88
Monotropism and Communication in Different Contexts 88
Individual Differences in Musical Preferences 90
References 92

7 Musical Journeys 2: Heightened Perceptual Experience and Musical Creativity 95
Sensory Disturbance in Autism 95
Auditory Hypersensitivity and Musical Development 97
Synaesthesia, Sensory Disturbance and Evolving Musicality 99
Creative Journeys 100
Auditory Perception Synaesthesia and Neurodiversity 101
James and Stéphane 103
References 107

8 Conclusions: Retrospective and Prospective	109
References	112
Index	115

1

Introduction to Autism and Music

Abstract Music is an important aspect of everyday life and a domain within which exceptional skills and talents can flourish. Researchers working in the field of autism science have traditionally explored musical strengths within the framework of domain-general cognitive models, and conclusions drawn from this work have influenced narratives about musicality in autism. In this introductory chapter I will outline these approaches to understanding musicality in autism in order to illustrate why a new approach is needed. The following chapters will present empirical and theoretical work on music and autism, before looking in detail at biographical accounts derived from extensive interviews conducted with autistic musicians working within different musical genres. The overarching aim of the book is to explore the complex relationship between autism and exceptional musicality in the context of evolving concepts of autism and the expert insights of autistic musicians.

Keywords Autism • Autism spectrum disorder • Musicality • Models of autism • Cognitive models of talent

Autism and Autism Spectrum Disorder

The earliest widely disseminated account of autism was published by Leo Kanner in 1943 [1]. Kanner, an Austrian-born psychiatrist and physician, was at the forefront of a newly emerging discipline of child psychiatry, founding a specialist clinic at the John Hopkins University in the US in 1930, and publishing the first English-language textbook on child psychiatry in 1935 [2]. In his seminal 1943 paper, 'Autistic Disturbances of Affective Contact', he documented the cases of 11 children, who at different points in their early lives had been referred to the clinic for 'peculiarities of behaviour' or for assessment for intellectual disability. Kanner and colleagues in his own and other institutions had been following the development of these children for several years, and his account reflects extensive observation, reports from the children's family members, and evolving ideas about the concept of autism. Within psychiatry the term 'autism' had been used to describe a person's separation from their reality and was most strongly associated with schizophrenia at that time. However, in his account of these children Kanner described 'autistic disturbances of affective contact' that were independent of delusions, hallucinations or other symptoms associated with schizophrenia. The children showed a complex constellation of behaviours and physiological symptoms that had not previously been conceptualised as a discrete syndrome in the psychiatry literature. In addition to what were conceptualised as interpersonal difficulties, the children showed an unusual profile of cognitive strengths and difficulties. Formal testing of several children previously deemed to be intellectually handicapped revealed average or strong intellectual skills. Within the group, examples of exceptional skills, for example in memory and vocabulary, were observed. Kanner noted that some of the children appeared to be intrigued by things that didn't usually capture the imagination of typically developing children, describing how these interests were 'governed by an anxiously obsessive desire for the maintenance of sameness'. In drawing together his observations, Kanner concluded that the children presented with 'extreme autism', which he defined as an 'innate inability to form the usual affective contact' and co-occurring repetitive ritualistic behaviours. Although there have been considerable shifts in the way autism has been conceptualised since Kanner's early description, the importance of his clinical insights continues to be acknowledged [3] and many of the characteristics he described are currently included in formal diagnostic criteria for autism.

Autism was first included in the *Diagnostic and Statistical Manual of Mental Disorders* (DSM-111) (American Psychiatric Association: APA) under the descriptive term 'infantile autism' in 1980 [4]. The criteria for this condition included a pervasive lack of responsiveness to other people, difficulties with language and atypical responses to the environment. It also specified that these 'symptoms' must be observed by 30 months of age and that delusions, hallucinations or other 'symptoms' of schizophrenia should not be in evidence. Autism was initially believed to be a very rare condition, with the first epidemiological study, carried out in the UK in 1966 [5], reporting a prevalence rate of 4.5 in 10,000, or one in 2220, 8–10-year-old children. In the 2013 revision of the DSM (DSM-5) [6] the broad umbrella term autism spectrum disorder (ASD) was adopted to include conditions like Asperger's syndrome and autistic disorder, and to reflect individual differences in levels of 'symptom' severity. Criteria in this edition specified that an individual must display 'persistent deficits in social communication and interaction' and demonstrate 'restricted, repetitive patterns of behaviour, interests, or activities'. It is currently estimated that one in 34 10–14-year-old children and adolescents living in the UK are diagnosed with ASD [7]. A similar estimate, of one in 36, has been reported in 8-year-old children living in the US [8]. Zeidan and colleagues (2022) [9] reported a worldwide autism prevalence rate of one in 100 with variance within and across countries reflecting differences in geographic, ethnic and sociodemographic factors. Increases in autism prevalence rates have been linked with a broadening of diagnostic criteria. For example, Polyak, Kubina and Girirajan (2015) [10] analysed data from children enrolled in special education provisions in the US during the period spanning 2000 to 2010 and reported that diagnostic recategorisation from intellectual disability to autism helped explain a 331% increase in autism prevalence in this population. In a study carried out in the UK, Russell and colleagues (2022) [11] analysed primary care data spanning 1998 to 2018 and reported a 787% rise in cases of autism, with the steepest increases amongst females, adult males and individuals who do not have intellectual disability. Autism is an exceptionally heterogeneous condition, not only in the severity and expression of 'symptoms', but also in the genetic and other biological and environmental factors that have been associated with it [12]. Chapter 3 will present a brief overview of changes in diagnostic criteria and findings from studies identifying factors that are associated with heterogeneity in autism.

Music, Musicality and Musical Expertise

The human propensity to listen to, create and perform music is both ancient and universal. However, music can adopt and serve seemingly limitless forms and functions, and this makes it very difficult to define. Davies (2021) [13] outlined and critically evaluated multiple definitions of music. For example, socio-cultural definitions of music place emphasis on the historical and social contexts within which it is created and experienced, whilst structural definitions focus more directly on musical materials, analysing the organisation of, and relationship between, the parts or elements within them. Definitions based on music's functions most powerfully illustrate its importance in everyday life. For example, Cross and Woodruff (2009) [14] have described how music adds significance to rites of passage that involve individuals and their communities, increases co-operation and affiliation within groups and communities, and marks the significance of an individual's relationships with their natural, social and/or political environment. Evolutionary theories of music (described in Chap. 2) stress the importance of music's communication and affiliative functions, arguing that the emergence of musicality was adaptively advantageous for our early ancestors. Given this, it is tempting to assume that the 'essence' of music lies in its communicative potential. However, music also serves important intra-personal functions. Davies (2012) [15] has described how it may be used to induce and sustain trance states, dissociation or inward focus, which serves to 'cocoon the individual in a wall of sound that excludes others and the world'. Moreover, empirical findings [16] (described in Chap. 2) suggest that for individuals living in contemporary Western societies, music's intra-personal functions may be more important than its interpersonal ones.

The formulation of a universally accepted definition of music may well be an unachievable goal. However, it is clear that music, in all its facets, continues to fascinate and delight our species, and this in itself demands exploration.

Honing, Ten Cate, Peretz and Trehub (2015) [17] have stressed the importance of distinguishing music and musicality, proposing that the latter may be defined as 'a natural, spontaneously developing trait based on and constrained by biology and cognition'. In common with all complex human traits, musicality shows considerable variation across individuals. Historically, debates about cases of exceptional musicality have been couched in the nature/nurture distinction [18], with proponents of these competing positions arguing for the primacy of 'innate' ability or for environmental factors such as 'deliberate practice' [19, 20]. However, more recent theorising about talent acknowledges the bi-directional influence of genes and environments and avoids oversimplistic explanations for its emergence. For example, the Multifactorial Gene-Environment Interaction Model (MGIM) developed by Ullén, Mosing and Hambrick (2017) [21] implicates biological, cognitive, personality, motivational and experiential factors and their interactions in the emergence of exceptional skills. Consistent with research showing that long-term deliberate practice results in anatomical and functional reorganisation of brain circuits and enhanced sensorimotor skills that are directly relevant to domains of expertise [20, 22], the model affirms the importance of practice. However, the choice of talent domain (e.g. music, football, chess) as well as the quality and quantity of practice an individual engages in are influenced by cognitive, personality, motivational and other factors. Within this model, an individual's domain general characteristics may facilitate some types of expertise or talent. For example, high levels of intelligence and working memory may be advantageous in the fields of music and chess, whilst strong sensory-motor control may facilitate achievements in sports and dance. Finally, the model proposes that genetic influences, both on approaches to practice and on traits that are advantageous within specific domains, will also be important. Theoretical advances in understanding exceptional skills and talents have important implications for autism science. Results from empirical studies suggest a high preponderance of exceptional skills in the domains of music, art, reading, mathematics and information technology in autistic people [23, 24]. However, approaches to understanding and studying these have not been influenced by advances in talent research but have more commonly tested hypotheses drawn from cognitive models of autism.

Cognitive Models of Autism

The theoretical model that has had the greatest impact, both on narratives around autism and on methods of studying social cognition is the Theory of Mind deficit account (ToM), first proposed by Baron-Cohen, Leslie and Frith in 1985 [25]. This paradigm marked a radical departure from traditional developmental approaches to studying cognition in children [26] and reflected both ongoing changes in the way autism was conceptualised and paradigm shifts within psychology. Historical analyses carried out by Evans (2013, 2017) [27, 28] and Verhoeff (2013) [29] have documented the way that early beliefs about autism, and its association with emotional disturbance and hallucinations, were abandoned as autism was increasingly conceptualised as a social and communication 'disorder'. Within psychology, the information processing paradigm, in which the computer serves as a metaphor for mental processing, had become dominant by the early 1980s and Fodor's 1983 book, *The Modularity of Mind* [30] introduced the concept of independent mental modules that operate on specific kinds of environmental inputs. In their first, highly influential paper describing the ToM deficit hypothesis, Baron-Cohen and colleagues reported on the results from a study comparing groups of children with autism, Down syndrome and typical development on a task that probed their understanding of false belief. In this task, children observe a sequence of events in which a story protagonist (Sally) places her marble in a basket before leaving the room. Whilst Sally is out of the room a second protagonist (Anne) moves the marble to another location. The children are then asked where Sally will look for her marble when she returns. The results from the experiment showed that most of the children in the Down syndrome and typically developing groups, but only 20% of the autistic children, believed that Sally would search for her marble in the basket. The authors proposed that this finding resulted from an autism-specific deficit in understanding that Sally's actions would be influenced by her belief about where the marble was.

The ToM deficit account rapidly gained traction amongst researchers working in autism science, and the concept was expanded to account for multiple aspects of social and emotional understanding and empathy. However, the influence of this account, pervasive as it was, has met with increasing criticism over the last two decades. For example, the expected outcome of research into developmental conditions is that it will provide a scientific base for the development of therapies that promote wellbeing and learning. However, therapeutic interventions aimed at teaching ToM skills have not resulted in sustained or generalisable benefits for autistic people [31]. Evidence has shown that difficulties on ToM tasks are not unique to or universal within autism. Poor performance on ToM tests has been reported in studies of individuals with Down syndrome [32], Williams syndrome [33], major depressive disorder [34], schizophrenia [35] and anorexia nervosa [36]. Moreover, research shows that many of the measures used to probe ToM skills do not correlate, and this raises questions about the construct validity of this paradigm [37–39]. An alternative approach has been to study autism within a neuro-developmental framework [40, 41]. These accounts, outlined in Chap. 3, place emphasis on sensory and other disturbances that influence the child's early and ongoing experiences within their physical and social environments. They describe developmental trajectories that are influenced by both biological and experiential factors and give rise to profiles of difficulties and strengths in autism. The most powerful challenge to the ToM deficit account of autism has emerged from research situated within the neurodiversity framework. These results show that 'social-communication difficulties' in autism cannot be explained in the context of 'mentalizing deficits' within individuals but rather in a context that considers interactions between autistic and non-autistic people [42]. This explanation has gained support from rigorously designed empirical studies and some of these will be described in Chap. 4.

Cognitive Models of Talent in Autism

In his 1943 article Kanner reported that some of the children he described were unable to 'experience wholes without full attention to the constituent parts'. He also considered how atypical configural processing might influence their acquisition of skills. Most prominent amongst more recent cognitive models seeking to explain skills and talents in autism are the Weak Central Coherence Theory (WCC), the Enhanced Perceptual Functioning theory (EPF), the Hyper-Systemising theory and the Veridical Mapping theory (VM). Although these models differ in detail, they each propose that autistic people differ in perceptual or cognitive style from non-autistic people and that this difference is linked with the emergence of exceptional skills in this group. Below, I will briefly outline each of these theories in turn.

In her original formulation of the WCC theory in 1989 [43], Frith hypothesised that the typically observed propensity to extract gist, or process information globally/in context, is disrupted in autism. Empirical studies carried out by Shah and Frith had shown superior performance in autistic children compared with typically developing children on the Children's Embedded Figures task [44] and the Block Design Task from the Weschler Intelligence Scales [45]. These tasks are used to measure a child's ability to locate simple shapes that are embedded in complex visual designs and to replicate a visual design with small, patterned blocks. In response to these and later experimental findings, Happe and Frith (2006) [46] proposed that cognitive strengths in autism result from a local or detail-focused information processing style associated with weak global processing. The first alternative explanation for autistic strengths on specific tasks was proposed by Mottron and Burack in 2001 [47], and later revised by Mottron, Dawson, Soulieres, Hubert and Burack in 2006 [48]. Although experimental studies comparing autistic and non-autistic groups had continued to show superior local or detailed-focused processing in the autism groups, several studies also reported 'intact' global processing on visual-spatial tasks [49, 50]. A similar finding emerged in research exploring music perception in autism. For example, early studies testing perception of chords and melodies revealed fine-grained pitch discrimination and holistic processing of these small-scale musical forms [51–53]. Moreover, studies using harmonic priming paradigms showed that autistic children's expectancies about upcoming chords were more powerfully influenced by global than by local harmonic contexts [54, 55]. In acknowledgement of these findings the proponents of the EPF theory did not link perceptual strengths with 'deficits' in higher-order cognitive processes. Instead, they proposed that autistic perception is less strongly and consistently dominated by the global processing bias that is characteristic in typical development. The most important aspect of the EPF theory is that it stresses the importance of perceptual information, proposing that for autistic individuals the physical qualities of stimuli may retain their salience, even when integrated into more complex configurations. Music is especially rich in perceptual information, and the potential implications of enhanced perception for musical development in autism will be explored in detail in Chap. 7.

The Hyper-Systematising and VM models outline alternative accounts of the way structural information within talent domains is acquired and organised in cognition in autism. Whilst many forms of music are rich in detail, their higher-order or structural components are nevertheless of considerable importance, and a facility in organising and encoding this information cannot be explained in the context of accounts proposing a local bias or enhanced perception. The Hyper-Systematising theory, initially developed by Baron-Cohen, Richler, Bisarya, Gurunathan and Wheelwright in 2003 [56], proposes that autistic skills and talents reflect strengths in the ability to analyse, build, understand and predict rules within what the authors conceptualise as 'rule-based' domains. Within this model, Hyper-Systematising is not associated with global processing difficulties, but rather with superior perceptual processing and reduced empathic responses. The VM Model, developed by Mottron, Bouvet, Bonnel, Samson, Burack, Dawson and Heaton (2013) [57], builds on the assumption, stated in the EPF account, that perceptual information has increased salience in autism. According to this model, streams of patterned information, which may be relatively independent, may also be mapped together in a way that becomes automatic and stable over time. This capacity for mapping information is not unique to autism, as it is observed when typically developing children associate patterns of auditory and visual information when learning to read. However, within the VM model, this mapping mechanism is utilised more widely and supports skill acquisition in domains of special interest in autism. In an empirical study testing this model, Bouvet, Donnadieu, Valdois, Caron, Dawson and Mottron (2014) [58] described the case of FC, an autistic man with synaesthesia, hyperlexia and absolute pitch.[1] In addition to measuring and validating FC's different skills, the authors questioned him about how he had acquired them. In describing his absolute pitch skills, FC recalled how he had consciously linked the seven tones of the diatonic scale (i.e. A, B, C, D, E, F, G) with the seven days of the week, and the 12 semitones in the chromatic scale (i.e. A, A#, B, C, C#, D, D#, E, F, F#, G, G#) with the 12 months of the year. When mapping tones in melodies that extended beyond a single octave, FC described how he mentally represented previous and future weeks and years in lower or higher octave transpositions. In reviewing the findings from their study, Bouvet and colleagues proposed that FC's skills resulted from a combination of perceptual learning and memory capacities.

Rethinking Thinking About Musicality in Autism

In Kanner's earliest account of autism, he not only described the children's difficulties but also noted their cognitive strengths and the intensity of their engagement in their personal interests. Later cognitive models of autism attempted to provide a scientific framework within which perceptual and cognitive mechanisms, believed to facilitate talent development, could be studied. For example, early experimental studies testing hypotheses drawn from the WCC and EPF theories identified strengths in pitch discrimination and memory [51, 59, 60] in autism, and the Hyper-Systematising and Veridical Mapping (VM) models propose

---

[1] Synaesthesia: stimulation in one sensory modality (e.g. hearing) leads to involuntary experiences in a second sensory modality (e.g. smell). Hyperlexia: precocious reading ability (e.g. at around 2–5 years).

Absolute Pitch: an ability to identify a musical tone without the need for a reference tone.

testable accounts of the processes that enable individuals to order and encode complex structural information within this domain. However, there are many fundamental questions about musicality in autism that cannot be addressed in the framework of these cognitive models. For example, empirical studies have shown that autistic children demonstrate good [61–63], or exceptionally good [64], identification of emotions depicted in music and very little is known about the strengths in performance, improvisation and composition, which will be described in the later chapters of this book. The underlying reason why this approach fails in the case of music is that it assumes that music can be conceptualised as a 'rule based' or 'closed system' that can be defined in purely structural terms [65]. This disregards music's innumerable functions and the powerful and ubiquitous effects it has on those who create and engage with it. This latter point will be elaborated in Chap. 2 where the nature and development of musicality will be explored within a bio-musicology framework.

Evidence from autism science shows that research is more relevant, rigorous and ethical when autistic people are closely involved in all aspects of the research process [66]. Work based on this insight is radically changing the way autism is conceptualised within science and within the wider community. Early autobiographical accounts, written by Temple Grandin [67–69], and Donna Williams [70, 71], paved the way for what is now a substantial body of work documenting the experiences of living with autism. In a recent article Dinishak and Akhtar (2023) [72] discussed the question of how perspectives drawn from autobiographical accounts can most effectively be integrated into autism science. Scholars working within the fields of autism, music psychology and ethnomusicology have already begun the task of studying musical experience in autism from a first-person perspective [73–76]. Some of this work will be described in Chap. 4. These studies exploring the roles that music plays in everyday life have provided powerful insights into the complexity and breadth of the participating individuals' musical experiences. Building on this research, the aim of this book is to investigate the emergence and nature of musical talents in autism from a first-person perspective and within the context of current research on autism and music.

References

1. Kanner, L. (1943). Autistic disturbances of affective contact. *Nervous Child, 2*, 217–250.
2. Bond, E. D. (1936). Child psychiatry. By Leo Kanner, MD (Springfield: Chas. C. Thomas, 1935.). *American Journal of Psychiatry, 93*(1), 240-a.
3. Rosen, N. E., Lord, C., & Volkmar, F. R. (2021). The diagnosis of autism: From Kanner to DSM-III to DSM-5 and beyond. *Journal of Autism and Developmental Disorders, 51*(12), 4253–4270. https://doi.org/10.1007/s10803-021-04904-1. Epub 2021 Feb 24. PMID: 33624215; PMCID: PMC8531066.
4. American Psychiatric Association. (1980). *Diagnostic and statistical manual of mental disorders* (3rd ed.). American Psychiatric Association.
5. Lotter, V. (1966). Epidemiology of autistic conditions in young children: 1. Prevalence. *Social Psychiatry, 1*, 124–135.
6. American Psychiatric Association (Ed.). (2013). *Diagnostic and statistical manual of mental disorders* (5th ed.). American Psychiatric Association.
7. O'Nions, E., Petersen, I., Buckman, J. E. J., Charlton, R., Cooper, C., Corbett, A., Happé, F., Manthorpe, J., Richards, M., Saunders, R., Zanker, C., Mandy, W., & Stott, J. (2023). Autism in England: Assessing underdiagnosis in a population-based cohort study of prospectively collected primary care data. *The Lancet Regional Health - Europe, 29*, 100626. https://doi.org/10.1016/j.lanepe.2023.100626. PMID: 37090088; PMCID: PMC10114511.
8. Maenner, M. J., Warren, Z., Williams, A. R., et al. (2023). Prevalence and characteristics of autism spectrum disorder among children aged 8 years — Autism and developmental disabilities monitoring network, 11 sites, United States, 2020. *MMWR Surveillance Summaries, 72*(SS-2), 1–14. https://doi.org/10.15585/mmwr.ss7202a1
9. Zeidan, J., Fombonne, E., Scorah, J., Ibrahim, A., Durkin, M. S., Saxena, S., Yusuf, A., Shih, A., & Elsabbagh, M. (2022). Global prevalence of autism: A systematic review update. *Autism Research, 15*(5), 778–790. https://doi.org/10.1002/aur.2696. Epub 2022 Mar 3. PMID: 35238171; PMCID: PMC9310578.
10. Polyak, A., Kubina, R. M., & Girirajan, S. (2015). Comorbidity of intellectual disability confounds ascertainment of autism: Implications for genetic diagnosis. *American Journal of Medical Genetics. Part B,*

Neuropsychiatric Genetics, 168(7), 600–608. https://doi.org/10.1002/ajmg.b.32338. Epub 2015 Jul 22. PMID: 26198689.
11. Russell, G., Stapley, S., Newlove-Delgado, T., Salmon, A., White, R., Warren, F., Pearson, A., & Ford, T. (2022). Time trends in autism diagnosis over 20 years: A UK population-based cohort study. *Journal of Child Psychology and Psychiatry, 63,* 674–682. https://doi.org/10.1111/jcpp.13505
12. Masi, A., DeMayo, M. M., Glozier, N., & Guastella, A. J. (2017). An overview of autism spectrum disorder, heterogeneity and treatment options. *Neuroscience Bulletin, 33*(2), 183–193. https://doi.org/10.1007/s12264-017-0100-y. Epub 2017 Feb 17. PMID: 28213805; PMCID: PMC5360849.
13. Davies, S. (2021). Evolution. In T. McAuley, N. Nielsen, & J. Levinson (Eds.), *Oxford handbook of music & philosophy* (pp. 677–703). Oxford University Press.
14. Cross, I., & Woodruff, G. E. (2009). Music as a communicative medium. In R. Botha & C. Knight (Eds.), *The prehistory of language* (pp. 77–99). Oxford University Press.
15. Davies, S. (2012). *The artful species: Aesthetics, art and evolution.* Oxford University Press.
16. Schäfer, T., Sedlmeier, P., Städtler, C., & Huron, D. (2013). The psychological functions of music listening. *Frontiers in Psychology, 4,* 511. https://doi.org/10.3389/fpsyg.2013.00511. PMID: 23964257; PMCID: PMC3741536.
17. Honing, H., ten Cate, C., Peretz, I., & Trehub, S. E. (2015). Without it no music: Cognition, biology and evolution of musicality. *Philosophical Transactions of the Royal Society of London. Series B, Biological Sciences, 370*(1664), 20140088. https://doi.org/10.1098/rstb.2014.0088. PMID: 25646511; PMCID: PMC4321129.
18. Howe, M. J. A., Davidson, J. W., & Sloboda, J. A. (1998). Innate talents: Reality or myth? *Behavioral and Brain Sciences, 21*(3), 399–407. ISSN 0140-525X. https://doi.org/10.1017/S0140525X9800123X
19. Ericsson, K. A., Krampe, R. T., & Tesch-Römer, C. (1993). The role of deliberate practice in the acquisition of expert performance. *Psychological Review, 100*(3), 363–406. https://doi.org/10.1037/0033-295X.100.3.363
20. Sloboda, J. A., Davidson, J. W., Howe, M. J. A., & Moore, D. G. (1996). The role of practice in the development of performing musicians. *British Journal of Psychology, 87*(2), 287–309. https://doi.org/10.1111/j.2044-8295.1996.tb02591.x

21. Ullén, F., Mosing, M. A., & Hambrick, D. Z. (2017). The multifactorial gene-environment interaction model (MGIM) of expert performance. In *The science of expertise* (pp. 365–375). Routledge.
22. Hyde, K. L., Lerch, J., Norton, A., Forgeard, M., Winner, E., & Evans, A. C. (2009). Musical training shapes structural brain development. *The Journal of Neuroscience, 29*, 3019–3025. others.
23. Howlin, P., Goode, S., Hutton, J., & Rutter, M. (2009). Savant skills in autism: Psychometric approaches and parental reports. *Philosophical Transactions of the Royal Society of London. Series B, Biological Sciences, 364*(1522), 1359–1367. https://doi.org/10.1098/rstb.2008.0328. PMID: 19528018; PMCID: PMC2677586.
24. Bennett, E., & Heaton, P. (2012). Is talent in autism spectrum disorders associated with a specific cognitive and behavioural phenotype? *Journal of Autism and Developmental Disorders, 42*(12), 2739–2753. https://doi.org/10.1007/s10803-012-1533-9. PMID: 22527706.
25. Baron-Cohen, S., Leslie, A. M., & Frith, U. (1985). Does the autistic child have a "theory of mind"? *Cognition, 21*(1), 37–46. https://doi.org/10.1016/0010-0277(85)90022-8. PMID: 2934210.
26. Karmiloff-Smith, A. (1992). *Beyond modularity: A developmental perspective on cognitive science*. The MIT Press.
27. Evans, B. (2013). How autism became autism: The radical transformation of a central concept of child development in Britain. *History of the Human Sciences, 26*(3), 3–31. https://doi.org/10.1177/0952695113484320
28. Evans, B. (2017). *The metamorphosis of autism: A history of child development in Britain*. Manchester University Press. PMID: 28654228.
29. Verhoeff, B. (2013). Autism in flux: A history of the concept from Leo Kanner to DSM-5. *History of Psychiatry, 24*(4), 442–458. https://doi.org/10.1177/0957154X13500584
30. Fodor, J. A. (1983). *Modularity of mind: An essay on faculty psychology*. The MIT Press.
31. Marraffa, C., & Araba, B. (2016). Social communication in autism spectrum disorder not improved by theory of mind interventions. *Journal of Paediatrics and Child Health, 52*(4), 461–463.
32. Yirmiya, N., Solomonica-Levi, D., Shulman, C., & Pilowsky, T. (1996). Theory of mind abilities in individuals with autism, down syndrome, and mental retardation of unknown etiology: The role of age and intelligence. *Journal of Child Psychology and Psychiatry, 37*(8), 1003–1014. https://doi.org/10.1111/j.1469-7610.1996.tb01497.x. PMID: 9119934.

33. Tager-Flusberg, H., & Sullivan, K. (2000). A componential view of theory of mind: Evidence from Williams syndrome. *Cognition, 76*(1), 59–90. https://doi.org/10.1016/s0010-0277(00)00069-x. PMID: 10822043.
34. Bora, E., & Berk, M. (2016). Theory of mind in major depressive disorder: A meta-analysis. *Journal of Affective Disorders, 191*, 49–55. https://doi.org/10.1016/j.jad.2015.11.023. Epub 2015 Nov 23. PMID: 26655114.
35. Martin Brüne, M. D. (2005). "Theory of mind" in schizophrenia: A review of the literature. *Schizophrenia Bulletin, 31*(1), 21–42. https://doi.org/10.1093/schbul/sbi002
36. Bora, E., & Köse, S. (2016). Meta-analysis of theory of mind in anorexia nervosa and bulimia nervosa: A specific İmpairment of cognitive perspective taking in anorexia nervosa? *The International Journal of Eating Disorders, 49*, 739–740. https://doi.org/10.1002/eat.22572
37. Gernsbacher, M. A., & Yergeau, M. (2019). Empirical failures of the claim that autistic people lack a theory of mind. *Archives of Scientific Psychology, 7*(1), 102–118. https://doi.org/10.1037/arc0000067
38. Klin, A. (2000). Attributing social meaning to ambiguous visual stimuli in higher-functioning autism and Asperger syndrome: The social attribution task. *Journal of Child Psychology and Psychiatry, 41*(7), 831–846. PMID: 11079426.
39. Klin, A., Jones, W., Schultz, R., & Volkmar, F. (2003). The enactive mind, or from actions to cognition: Lessons from autism. *Philosophical Transactions of the Royal Society of London. Series B, Biological Sciences, 358*(1430), 345–360. https://doi.org/10.1098/rstb.2002.1202. PMID: 12639332; PMCID: PMC1693114.
40. Johnson, M. H. (2017). Autism as an adaptive common variant pathway for human brain development. *Developmental Cognitive Neuroscience, 25*, 5. https://doi.org/10.1016/j.dcn.2017.02.004
41. Gliga, T., Jones, E. J., Bedford, R., Charman, T., & Johnson, M. H. (2014). From early markers to neuro-developmental mechanisms of autism. *Developmental Review: Dr., 34*, 189–207. https://doi.org/10.1016/J.Dr.2014.05.003
42. Milton, D. E. (2012). On the ontological status of autism: The 'double empathy problem'. *Disability & Society, 27*(6), 883–887.
43. Frith, U. (2003). *Autism: Explaining the enigma* (2nd ed.). Blackwell Publishing.

44. Shah, A., & Frith, U. (1983). An islet of ability in autistic children: A research note. *Child Psychology & Psychiatry & Allied Disciplines, 24*(4), 613–620. https://doi.org/10.1111/j.1469-7610.1983.tb00137.x
45. Shah, A., & Frith, U. (1993). Why do autistic individuals show superior performance on the block design task? *Child Psychology & Psychiatry & Allied Disciplines, 34*(8), 1351–1364. https://doi.org/10.1111/j.1469-7610.1993.tb02095.x
46. Happé, F., & Frith, U. (2006). The weak coherence account: Detail-focused cognitive style in autism spectrum disorders. *Journal of Autism and Developmental Disorders, 36*(1), 5–25. https://doi.org/10.1007/s10803-005-0039-0. PMID: 16450045.
47. Mottron, L., & Burack, J. A. (2001). Enhanced perceptual functioning in the development of autism. In J. A. Burack, T. Charman, N. Yirmiya, & P. R. Zelazo (Eds.), *The development of autism: Perspectives from theory and research* (pp. 131–148). Lawrence Erlbaum Associates Publishers.
48. Mottron, L., & Burack, J. (Eds.). (2006). Autism: A different perception [Editorial]. *Journal of Autism and Developmental Disorders, 36*(1), 1–3. https://doi.org/10.1007/s10803-005-0048-z
49. Plaisted, K., Swettenham, J., & Rees, L. (1999). Children with autism show local precedence in a divided attention task and global precedence in a selective attention task. *Journal of Child Psychology and Psychiatry, and Allied Disciplines, 40*(5), 733–742.
50. Mottron, L., Burack, J. A., Iarocci, G., Belleville, S., & Enns, J. T. (2003). Locally oriented perception with intact global processing among adolescents with high-functioning autism: Evidence from multiple paradigms. *Journal of Child Psychology and Psychiatry, and Allied Disciplines, 44*(6), 904–913.
51. Heaton, P. (2003). Pitch memory, labelling and disembedding in autism. *Journal of Child Psychology and Psychiatry, 44*, 543–551. https://doi.org/10.1111/1469-7610.00143
52. Heaton, P. (2005). Interval and contour processing in autism. *Journal of Autism and Developmental Disorders, 35*(6), 787–793. https://doi.org/10.1007/s10803-005-0024-7. PMID: 16283085.
53. Jamey, K., Foster, N. E. V., Sharda, M., Tuerk, C., Nadig, A., & Hyde, K. L. (2019). Evidence for intact melodic and rhythmic perception in children with autism spectrum disorder. *Research in Autism Spectrum Disorders, 64*, 1–12, ISSN 1750-9467, https://www.sciencedirect.com/science/article/pii/S1750946719300480. https://doi.org/10.1016/j.rasd.2018.11.013

54. Heaton, P., Williams, K., Cummins, O., & Happé, F. G. (2007). Beyond perception: Musical representation and on-line processing in autism. *Journal of Autism and Developmental Disorders, 37*(7), 1355–1360. https://doi.org/10.1007/s10803-006-0283-y. Epub 2006 Dec 5. PMID: 17146705.
55. DePape, A. M. R., Hall, G. B., Tillmann, B., & Trainor, L. J. (2012). *Auditory processing in high-functioning adolescents with autism spectrum disorder* (Vol. 7, p. e44084).
56. Baron-Cohen, S., Richler, J., Bisarya, D., Gurunathan, N., & Wheelwright, S. (2003). The Systemising Quotient (SQ): An investigation of adults with Asperger syndrome or high functioning autism and normal sex differences. *Philosophical Transactions of the Royal Society, 358*, 361–374.
57. Mottron, L., Bouvet, L., Bonnel, A., Samson, F., Burack, J. A., Dawson, M., & Heaton, P. (2013). Veridical mapping in the development of exceptional autistic abilities. *Neuroscience and Biobehavioral Reviews, 37*(2), 209–228. https://doi.org/10.1016/j.neubiorev.2012.11.016. Epub 2012 Dec 5. PMID: 23219745.
58. Bouvet, L., Donnadieu, S., Valdois, S., Caron, C., Dawson, M., & Mottron, L. (2014). Veridical mapping in savant abilities, absolute pitch, and synesthesia: An autism case study. *Frontiers in Psychology, 5*, 106. https://doi.org/10.3389/fpsyg.2014.00106. PMID: 24600416; PMCID: PMC3927080.
59. Heaton, P., Hermelin, B., & Pring, L. (1998). Autism and pitch processing: A precursor for savant musical ability? *Music Perception: An Interdisciplinary Journal, 15*(3), 291–305. https://doi.org/10.2307/40285769
60. Bonnel, A., Mottron, L., Peretz, I., Trudel, M., Gallun, E., & Bonnel, A. M. (2003). Enhanced pitch sensitivity in individuals with autism: A signal detection analysis. *Journal of Cognitive Neuroscience, 15*(2), 226–235. https://doi.org/10.1162/089892903321208169. PMID: 12676060.
61. Heaton, P., Hermelin, B., & Pring, L. (1999). Can children with autistic spectrum disorders perceive affect in music?An experimental investigation. *Psychological Medicine, 29*(6), 1405–1410. https://doi.org/10.1017/s0033291799001221. PMID: 10616946.
62. Heaton, P., Allen, R., Williams, K., Cummins, O., & Happé, F. (2008). Do social and cognitive deficits curtail musical understanding? Evidence from autism and Down syndrome. *British Journal of Developmental Psychology, 26*(2), 171–182. https://doi.org/10.1348/026151007X206776
63. Quintin, E. M., Bhatara, A., Poissant, H., Fombonne, E., & Levitin, D. J. (2011). Emotion perception in music in high-functioning adolescents

with autism spectrum disorders. *Journal of Autism and Developmental Disorders, 41*(9), 1240–1255. https://doi.org/10.1007/s10803-010-1146-0. PMID: 21181251.
64. Sivathasan, S., Dahary, H., Burack, J. A., & Quintin, E.-M. (2023). Basic emotion recognition of children on the autism spectrum is enhanced in music and typical for faces and voices. *PLoS One, 18*(1), e0279002. https://doi.org/10.1371/journal.pone.0279002
65. Davis, S. (2012). On defining Music. *The Monist, 95*(4), 535–555.
66. Fletcher-Watson, S., Adams, J., Brook, K., Charman, T., Crane, L., Cusack, J., Leekam, S., Milton, D., Parr, J. R., & Pellicano, E. (2019). Making the future together: Shaping autism research through meaningful participation. *Autism, 23*, 943–953. https://doi.org/10.1177/1362361318786721
67. Grandin, T. (1991). *Emergence: Labeled autistic* (with Margaret Scariano, 1986, updated 1991), ISBN 0-446-67182-7.
68. Grandin, T. (1995). The learning style of people with autism: An autobiography. In *Teaching children with autism: Strategies to enhance communication and socialization.* Quill. ISBN 0-8273-6269-2.
69. Grandin, T. (2008). *The way I see it: A personal look at autism and Asperger's.* ISBN 9781932565720.
70. Williams, D. (1992). *Nobody nowhere/Donna Williams.* Doubleday.
71. Williams, D. (1994). *Somebody somewhere/Donna Williams.* Doubleday.
72. Dinishak, J., & Akhtar, N. (2023). Integrating autistic perspectives into autism science: A role for autistic autobiographies. *Autism, 27*(3), 578–587. https://doi.org/10.1177/13623613221123731
73. Allen, R., Hill, E., & Heaton, P. (2009). The subjective experience of music in autism spectrum disorder. *Annals of the New York Academy of Sciences, 1169*, 326–331. https://doi.org/10.1111/j.1749-6632.2009.04772.x. PMID: 19673801.
74. Korošec, K., Osika, W., & Bojner-Horwitz, E. (2022). "It is more important than food sometimes"; meanings and functions of music in the lives of autistic adults through a hermeneutic-phenomenological Lense. *Journal of Autism and Developmental Disorders, 54*, 366. https://doi.org/10.1007/s10803-022-05799-2
75. Matsuno, M., Auzenne, D., & Chukoskie, L. (2021). "All bets are off": Flexible engagement with music-listening technologies by autistic adults. *Psychology of Music, 49*(6), 1573–1588. https://doi.org/10.1177/0305735620971037
76. Bakan, M. B. (2018). *Speaking for ourselves: Conversations on life, music, and autism.* Oxford University Press.

2

Conceptualising Musicality as a Complex Human Trait

Abstract In Chap. 1 it was argued that the co-occurrence of autism and exceptional musicality cannot be legitimately explored in a framework that conceptualises music as a 'rule based' or 'closed' system of structured sounds, and musicality as a restricted set of perceptual, cognitive and memory skills. In this chapter, current thinking about music and musicality will be explored in the context of a biomusicology framework that conceptualises musicality as a species-typical trait of humans that may be studied using similar methods to those used to explore complex behaviours in other species.

Keywords Biomusicology • Causation • Mechanism • Ontogeny • Social relatedness

Musicality and the Biomusicology Framework

Fitch (2015) [1] has outlined four interconnecting principles guiding the study of musicality within a biomusicology framework. First, musicality should be defined as a behaviour that encompasses multiple components.

In his writings on music, Fitch has emphasised the importance of song, drumming, social syncronisation and dance, pointing out that these are culturally widespread and pervasive human behaviours. The second and third principles outlined by Fitch stress the importance of understanding musicality outside of the context of our own species and culture. Studies exploring musicality from a comparative perspective have tested musical components, such as relative pitch perception and/or the ability to synchronise to a musical beat, in different species [2]. This work aims to determine whether these characteristics are observed across a wide or narrow range of species or are only observed in humans. An early and highly influential hypothesis, outlined by Patel in 2006 [3], proposed that a propensity to move in time with an auditory beat in a precise, predictive and tempo-flexible way was a uniquely human characteristic that developed from a capacity for complex vocal learning. Whilst this hypothesis received considerable support, later evidence identifying syncronisation to a beat in a sulphur-crested cockatoo (*Cacatua galeritaeleonora*) [3] and vocal learning in whales, seals, elephants and bats [4] led to a revision of the hypothesis [5]. According to this new hypothesis, the advanced capacity for vocal learning seen in humans evolved, via the process of gene-culture co-evolution, into a neural adaptation for sustained beat perception and synchronisation.

The work of ethnomusicologists has provided rich cross-cultural data on different aspects of music and musicality. For example, the extensive programme of work carried out by Nettle [6, 7] enabled him to formulate ideas about commonalities in musical forms and functions across different cultures. Nettle described how all cultures have singing and some form of instrumental music, and that music serves important tradition-carrying functions. He also reported that music from diverse cultures frequently shares structural characteristics including, for example, the extensive use of repetition and variation in musical sequences, the use of the major second interval in melodic progressions, and a rhythmic structure based on a distinction among note lengths. Nettle also considered commonalities in the way that music is conceptualised across cultures and the settings within which musical activity occurs. For instance, music is widely conceived as a 'distinct unit of creativity' and, in association with dance and speech, frequently occurs within religious and/or ritual settings. Cross-cultural commonalities in music and musical behaviours have also been studied using quantitative methods. Savage, Brown, Sakai and Currie, (2015) [8] analysed 304 musical recordings from Africa, North and South America, South, East, and South East Asia, the Middle East, Europe and Oceana, and identified a number of characteristics that occurred with very high frequency across the corpus. These included some of the structural and other characteristics described by Nettle, but also a complex network of characteristics: repetitive formal structures, regular rhythms, simple syllabic singing style and the use of percussion instruments. This activity occurred within communal and participatory social contexts involving group performance and dancing. More recently, Mehr and colleagues (2019) [9] studied a corpus of ethnographic texts on musical behaviours from different cultures and reported that music varies along the dimensions of formality, arousal and religiosity, and is associated with infant care, healing, dance and love. Fitch's fourth principle of biomusicology is that research into musicality should adopt the ethological framework for understanding behaviour formulated by Tinbergen (1963) [10]. This framework comprises four questions, each probing different, but interrelated aspects of the behaviour under investigation. Of Tinbergen's four questions, two are proximate or 'how?' questions about how a behaviour works (causation or mechanisms) and how it develops over an individual's lifetime (ontogeny or development). The remaining two questions are ultimate or 'why?' questions about the evolutionary origins of a behaviour (evolution or phylogeny) and its adaptive value (function or adaptation). This approach to understanding musicality draws together insights from a spectrum of separate disciplines including comparative, developmental and evolutionary psychology, anthropology and ethology.

Causation or Mechanism

Causation or mechanism addresses the question of how a behaviour works. This level of explanation may be extraordinarily broad, encompassing molecular mechanisms, physical morphology, and other biological and behavioural factors. Typically it has been used to seek explanations for behaviours that enable animals to avoid predation, or the seasonal onset of mating behaviour and/or birdsong. Research investigating mechanism in songbirds has identified genetic, neural and sensory response processes that, together with what has been learned from research exploring ontology, evolution and function, have elucidated the factors that elicit song learning and production in these species [11]. The task of identifying causation in musical behaviour is hindered by the varying and multifaceted nature of its forms and the related problem of how these can be encapsulated in a single definition. This caveat in no way suggests that birdsong and other animal behaviours studied in the context of Tinbergen's four questions are not complex. Rather, it suggests that human musicality may reach an exceptionally high order of variability and complexity which will in turn be reflected in mechanism or causation. A second caveat in applying this level of analysis to the study of musicality is that musical behaviours show considerable increases in complexity over the lifespan and causation can be difficult to differentiate from development or ontology. In an appraisal of Tinbergen's four questions, Bateman and Laland (2012) [11] discuss this problem and propose that the term 'mechanisms of control' suggests a more specific focus on the links between current behaviours and mechanism. Scientists working in the fields of neuroscience and music psychology have made considerable progress in mapping the cognitive and brain correlates of musical behaviours [12]. This includes, for example, mechanisms linked with the experience of 'chills' [13, 14] and 'anticipation' during music listening [15]. This approach to studying musicality has been important for evaluating theoretical models of music listening [16] and continues to provide a scientific knowledge base for the development of music-based therapies [17, 18].

Ontogeny or Development

In humans, conception and development occur within rich social environments that in interaction with biological inheritance continuously shape development over the lifespan [19]. Music is a prominent element in most social environments, and developmental psychologists working in the field of music psychology have carried out extensive research into musical responses in infants [20]. This work has identified a powerful early interest in infant-directed song [21] and shown that this form of communication is more effective than infant-directed speech in capturing attention [22], and in delaying the onset of distress [23]. From around five months infants move more in response to music and other rhythmically regular sounds than to speech. Moreover, the infant's degree of rhythmic co-ordination with the sounds they hear is positively related to their displays of positive emotion [24]. Cirelli, Einarson and Trainor (2014) [25] reported that by 14 months, infants who have shared a musical beat with another person show increased pro-social behaviour towards that person. In a very recent study, Lense, Schultz, Astésano and Jones (2022) [26] used eye tracking technology to test whether infants viewing audio-visual recordings of naturalistic nursery songs would synchronise their looking patterns with the rhythms of the songs. The results showed that this effect was observable by two months and progressively strengthened in groups of 4- and 6-month-old infants. The authors proposed that the rhythm of infant-directing singing attunes infants to precisely timed social-communicative information and supports social learning and development. In addition to identifying early social-emotional responses to music, researchers have identified perceptual and cognitive competencies that enable infants to learn about the characteristics of the music they hear around them. For example, studies have shown that infants are sensitive to the structural components of scale and melody [27–30] and the temporal and rhythmic components of music [31, 32]. Moreover, this information is retained in short- and long-term memory [33–35].

In the West, theorising about age-related change in cognitive skills has been strongly influenced by the stage theory of Jean Piaget (1896–1980). Piaget stressed the importance of the child's exploration and mastery of their environment. Exploration exposes the child to new information and this brings about cognitive change via the action of two interacting processes. The first of these is assimilation where newly encountered information can fit into a pre-existing schema. The second process, accommodation, occurs when new information cannot be assimilated into an existing schema, and the schema is modified. Based on an analysis of over 700 musical compositions, obtained from 3–15-year-old children and adolescents, Swanick and Tillman (1986) [36] formulated a developmental model that contextualises the child's emerging musicality, in the interplay between cognitive development, creativity and interactions within their cultural niche. According to this model, modes of engagement with music are initially characterised by *mastery* (0–4 years). Infants are very interested in sound, and whilst sound production is not organised, they achieve increasing mastery over the means of making sounds during this phase. The authors describe how young children who are learning about animals imitate the sounds they make, and in doing so accommodate a new understanding of the unique characteristics of the animals they are imitating. In a similar way, imitating musical phrases enables them to accommodate the structural and expressive characteristics of the music they hear in their environment. The *imitation phase* (4–9 years) marks a rise in personal expressivity in song production, and from around five years the child begins to demonstrate an internalisation of musical conventions in melodic and rhythmic patterning. In the third *imaginative play phase* (10–15 years), mastery of musical conventions and technical skills becomes increasingly consistent and children begin to engage in imaginative musical deviation. In their model, Swanick and Tillman link imaginative play with the process of assimilation where children are able to go beyond rules and generate their own forms. From around 13 to 14 years, children begin to identify with recognisable musical communities, and the final *metacognitive stage* (15+ years) marks the point at which an individual is able to theorise about musical styles and music's expressive qualities. Research into the ontogeny of music has provided some of the most powerful insights into this most complex of human traits. Empirical studies of infants have revealed an early readiness to attend to, learn about and respond to music, and Swanick and Tillman's model conceptualises the unfolding of musicality during childhood and adolescence as a complex interplay between these early predispositions, maturation, cultural heritage and individual differences.

Evolution

Evolution, or phylogeny, addresses questions about which ancestors first possessed a particular trait, what the antecedents to the trait are, and what selective pressures shaped it in the past. As an ethologist working on the evolution of behaviour, Tinbergen's thinking was strongly influenced by Darwin's theory of natural selection. According to this theory, natural selection occurs when individuals possess heritable traits that increase their own and their descendants' reproductive success. One example of a heritable trait is dense fur, which for animals living in a cold climate reduces the amount of energy needed to keep warm, thereby increasing their chances of survival and opportunities for reproduction. When considering the antecedents of musicality it is important to consider how traits, or 'adaptations', change over evolutionary time. In Darwin's sixth and final edition of the *Origin of Species* (1872) [37], he described how traits may have evolved from earlier traits that served different functions—a process later described by Gould and Vrba (1982) [38] as exaptation. Probably the most highly cited example of an exaptation is feathers, which are believed to have first evolved for insulation in dinosaurs and now serve the function of enabling flight in birds [39]. Within the biomusicology framework musicality is conceptualised as a complex trait, comprising multiple components, and there has been considerable debate about whether different components of musicality evolved for specifically musical functions or were exaptations of other traits [40]. The quest to understand the evolutionary origins of musicality was given considerable impetus by Pinker (1997) [41], who proposed that music is not an adaptation but a product of human culture (termed technology) and that the different components of musicality had evolved to serve other functions. These adaptations included motor control and other behaviours that enabled individuals to interpret and respond to the signals (e.g. prosodic, emotional calls) and physical characteristics (auditory scene analysis and habitat selection) they encountered in their everyday environments. More recent theoretical accounts by Patel (2018) [42] and Shilton (2022) [43] also conceptualise music as a product of human culture or technology. For example, Shilton has proposed that musicality emerged after the unique features of human sociality had been established. However, in this account these behavioural developments were later accommodated genetically, and musicality is conceptualised as the product of gene-culture co-evolution.

Theorising about the evolutionary origins of music has been increasingly influenced by scientific advances in understanding the factors that determine inheritance. Niche construction theory [44] describes the modifications that animals make to their environments and the effects these modifications have on the processes of evolution. Niche construction generates ecological legacies, which serve to alter the environmental conditions within which individuals are conceived, born and develop. The activities and learning that takes place within these ecological niches then influences evolution by modifying patterns of natural selection. This process was succinctly described by Levins and Lewontin (1985) [45] who proposed that 'the organism influences its own evolution, by being both the object of natural selection and the creator of the conditions of that selection' (p 106). Around 1.7 to 2.0 million years ago, humans first controlled fire, and this important cultural innovation is believed to have driven both biological and cultural change in early humans. Fire enabled early humans to cook meat and other foods and has been associated with increases in brain size and changes in other physiological characteristics [46]. Fire building also brought about changes in the environmental conditions within which cultural adaptations could occur. Shilton (2022) has linked four early cultural developments with musicality in humans. These include the evolution of a technological niche in which increasing complexity in tool production co-evolved alongside changes in cognitive, affective and physiological capacities. Shared intentionality, arising in the context of tool making, hunting and other forms of co-operative behaviour, increased biobehavioural synchrony, evidenced in bodily movement, eye gaze and attention following. The emergence of complex social structures, for example kinship groups and multilevel societies, also gave rise to new forms of interaction and joint action. Multilevel societies, in which small units of individuals were nested within a larger social matrix, provided new platforms for the development of complex social behaviours. Shilton describes how music would have been a viable strategy for extending the duration and intensity of group-co-ordinated activities such as coalition displays. Moreover, these different social groupings provided an ideal platform for the ritualisation of these entrainment-based interactions.

Several prominent theoretical accounts have linked the emergence of music with specific developments in social behaviours in early humans. Within the biomusicology framework, song is a core component of musicality, and the evolutionary psychologist Robin Dunbar (2017) [47] has written extensively about adaptations that drove the evolution of complex vocal behaviours in early humans. Group living is advantageous for most primates and for humans it provided protection against predators, and enabled new forms of co-operation, seen, for example, in hunting, farming, herding and childcare. However, group living presents challenges and is only characteristic in species that have evolved mechanisms for promoting within-group affiliation and conflict resolution. Dunbar describes how some species of non-human primates maintain positive group relations by engaging in regular one-to-one grooming (i.e. close inspection and cleaning of fur and skin). This behaviour activates neuroendocrine mechanisms associated with bonding and regulates behavioural responses between individuals. However, it is time-consuming and places constraints on group size. According to Dunbar, the benefits of living in expanded groups drove the search for a mechanism that could overcome this constraint and trigger these neuroendocrine responses in greater numbers of individuals at the same time. The solution to this problem came in the form of wordless chorusing, which over time evolved into music and language.[1]

In a second account Dissanayake (2000, 2009) [48, 49] proposed that human musicality is rooted in the early emergence of affective, communicative and affiliative interactions between mothers and infants. Pair-bonding behaviours are observed in many mammalian species, though the mechanisms of these behaviours show considerable differences in complexity across species [50]. Anatomical developments that resulted in

---

[1] In his Musilanguage model Brown (1999) [57] describes an early stage in human development during which a single form of communicative vocalisation emerges as two separate systems dedicated to music and language. Whilst these new systems shared some structural and psychoacoustic characteristics, music became increasingly specialised for expressing emotion whilst language became increasingly referential.

increased brain size and short periods of gestation in *Homo erectus*, around two million years ago, created unique conditions for child rearing in humans. Offspring now had a protracted period of post-natal immaturity and this resulted in increased complexity in mother–infant bonding behaviours. According to Dissanayake, affiliative and communicative signals, used outside the context of child rearing, took on new perceptual, temporal and affective characteristics within this context. Moreover, infants evolved to preferentially attend and respond to these new modes of communication. Shilton discussed the importance of infant-directed communication in the context of kinship groups. He theorised that childcare would have become a distributed task within the group, and infant-directed communication, characterised by fine-grained behavioural synchrony, and associated with bonding, would have increasingly influenced modes of communication amongst adults.

A third theory, proposed by Miller (2000) [51], drew on Darwin's courtship hypothesis for birdsong and human music (1871) [52]. In his account Miller highlighted parallels between animal mating displays and virtuosic musical performances that provide opportunities for showcasing desirable male traits to potential female partners. Two very recent accounts, by Mehr and colleagues (2021) [53] and Savage, Loui, Tarr, Schachner, Glowacki, Mithen and Fitch (2021) [54], build on and develop these earlier ideas about the potentially adaptive role social and communicative behaviours played in the evolution of music. Mehr and colleagues propose that musicality evolved as a 'credible signal' serving dual adaptive functions. The first of these, infant-directed song, signalled parental attention to infants, whilst the second, observed in coordinated, entrained rhythmic displays, signalled coalition strength and size to observers. In the second account by Savage and colleagues it is proposed that social bonding, observed in the contexts of infant care, mate selection, group cohesion and coalition signalling, is the overarching function that drove the evolution of music.

Function or Adaptation

Within Tinbergen's framework the evolutionary history of a behaviour and its functions are clearly differentiated. Bateson and Laland (2013) [11] suggest the use of the term 'current utility' to emphasise this distinction and to avoid making assumptions about the evolutionary processes that generated a behaviour's functionality. The human quest to understand music's functions stretches far back into antiquity, with questions and conclusions reflecting the intellectual preoccupations (e.g. philosophical, theological, biological) of the times. Whilst scholars working within multiple disciplines continue to generate hypotheses about the utility of music, this question has also been addressed in empirical studies. For example, Sedlmeier, Städtler and Huron (2013) [55] probed questions about the functions of music listening in a survey study that included 834 children and adults (8–85 years). In order to determine which questions should be asked the authors reviewed a large corpus of theoretical and empirical literature on the proposed functions of music. This identified more than 500 items which appeared to cluster in social, emotional, cognitive and arousal-related function categories. As many of the functions were the same or similar, these were rationalised into a set of 129 items that were then included in the survey. The analysis of the data identified three distinct dimensions that were consistent over gender and age groups. The first and most important function was that music was used to *regulate arousal and mood*. Here, participants endorsed the importance of music for enhancing mood, for increasing motivation and physiological arousal, and as a valued and diverting pastime. The second most important dimension was *self-awareness*, which most strongly captured the participant's personal relationship with music. Participants reported that they listened to music because it enabled them to minimise the distractions from the outside world and focus on their own thoughts, emotions and identity. The final dimension *social relatedness* included items such as 'I listen to music because'… (a) 'it makes me belong', (b) 'it makes me feel connected to the world' or (c) 'it is a social experience'. Although this factor was statistically significant, the analysis showed that it motivated music listening to a significantly lesser degree than the other two factors. The findings from this study showed that whilst music's social and communication functions continue to be important, music's intrapersonal functions currently motivate music listening more powerfully.

Summary

Studying music in the framework proposed by Fitch enables the examination of all aspects of this complex phenomenon within a single inclusive paradigm. Evidence from comparative studies has shown that animals exhibit behaviours that may be conceptualised as components of musicality, and this has informed debates about the evolution and nature of musicality as a human trait [56]. Ethnomusicologists have documented the universal nature of music and musicality and provided important insights into cross-cultural similarities and differences in musical form and functions. The four questions proposed by Timbergen provide powerful tools for exploring the different facets of musicality. For ecologists an ultimate aim is to integrate what has been learned from exploration of the four questions in an overarching account of the behaviour under study. This presents an ambitious goal for scholars studying music. From the perspective of establishing an ecological framework for understanding musicality in autism, this approach highlights the inadequacy of 'rule-based' or 'closed system' conceptualisations of what music is. Musical traits are interwoven in the evolutionary history of our species and continue to serve multiple and important functions in our everyday lives.

References

1. Fitch, W. T. (2015). Four principles of bio-musicology. *Philosophical Transactions of the Royal Society of London. Series B, Biological Sciences, 370*(1664), 20140091. https://doi.org/10.1098/rstb.2014.0091. PMID: 25646514; PMCID: PMC4321132.
2. Patel, A. D., & Demorest, S. M. (2013). Comparative music cognition: Cross-species and cross-cultural studies. In D. Deutsch (Ed.), *The psychology*

of music (pp. 647–681). Elsevier Academic Press. https://doi.org/10.1016/B978-0-12-381460-9.00016-X

3. Patel, A. D., Iversen, J. R., Bregman, M. R., & Schulz, I. (2009). Experimental evidence for synchronization to a musical beat in a nonhuman animal. *Current Biology, 19*, 827–830. https://doi.org/10.1016/j.cub.2009.03.038
4. Janik, V. M., & Knörnschild, M. (2021). Vocal production learning in mammals revisited. *Philosophical Transactions of the Royal Society of London. Series B, Biological Sciences, 376*(1836), 20200244. https://doi.org/10.1098/rstb.2020.0244. Epub 2021 Sep 6. PMID: 34482736; PMCID: PMC8419569.
5. Patel, A. D. (2021). Vocal learning as a preadaptation for the evolution of human beat perception and synchronization. *Philosophical Transactions of the Royal Society B, 376*, 20211011.
6. Nettle, B. (1983). *The study of ethnomusicology: Twenty-nine issues and concepts*. University of Illinois Press.
7. Nettle, B. (2020) An ethnomusicologist contemplates universals in musical sounds and musical cultures. In N. Esllin, B. Merker, and S. Brown (eds). The origins of music. : The MIT Press, 463–472.
8. Savage, P. E., Brown, S., Sakai, E., & Currie, T. E. (2015). Statistical universals reveal the structures and functions of human music. *Proceedings of the National Academy of Sciences of the United States of America, 112*, 8987–8992.
9. Mehr, S. A., et al. (2019). Universality and diversity in human song. *Science, 366*(6468), eaax0868.
10. Tinbergen, N. (2005). On aims and methods of ethology. *Animal Biology, 55*(4), 297–321. https://doi.org/10.1163/157075605774840941
11. Bateson, P., & Laland, K. N. (2013). Tinbergen's four questions: An appreciation and an update. *Trends in Ecology & Evolution, 28*(12), 712–718, ISSN 0169-5347,. https://doi.org/10.1016/j.tree.2013.09.013
12. Zattore, R. (2023). *From perception to pleasure: The neuroscience of music and why we love it*. Oxford University Press. https://doi.org/10.1093/oso/9789197558287.001.0001
13. Blood, A. J., & Zatorre, R. J. (2001). Intensely pleasurable responses to music correlate with activity in brain regions implicated in reward and emotion. *Proceedings. National Academy of Sciences. United States of America, 98*, 11818–11823. https://doi.org/10.1073/pnas.191355898
14. Blood, A. J., Zatorre, R. J., Bermudez, P., & Evans, A. C. (1999). Emotional responses to pleasant and unpleasant music correlate with activity in

paralimbic brain regions. *Nature Neuroscience, 2*, 382–387. https://doi.org/10.1038/7299

15. Salimpoor, V., Benovoy, M., Larcher, K., et al. (2011). Anatomically distinct dopamine release during anticipation and experience of peak emotion to music. *Nature Neuroscience, 14*, 257–262. https://doi.org/10.1038/nn.2726
16. Huron, D. (2006). *Sweet anticipation: Music and the psychology of expectation*. The MIT Press.
17. Schlaug, G., Altenmüller, E., & Thaut, M. (2010). Music listening and music making in the treatment of neurological disorders and impairments [Editorial]. *Music Perception, 27*(4), 249–250. https://doi.org/10.1525/mp.2010.27.4.249
18. Thaut, M. H., McIntosh, G. C., & Hoemberg, V. (2015). Neurobiological foundations of neurologic music therapy: Rhythmic entrainment and the motor system. *Frontiers in Psychology, 5*, 1185. https://doi.org/10.3389/fpsyg.2014.01185. PMID: 25774137; PMCID: PMC4344110.
19. Odling-Smee, J., & Laland, K. N. (2011). Ecological inheritance and cultural inheritance: What are they and how do they differ? *Biological Theory, 6*, 220–230. https://doi.org/10.1007/s13752-012-0030-x
20. Trainor, L. J. (1996). Infant preferences for infant-directed versus noninfant-directed playsongs and lullabies. *Infant Behavior and Development, 19*, 83–92.
21. Trehub, S. E. (2001). Musical predispositions in infancy. *Annals of the New York Academy of Sciences, 930*, 1–16.
22. Nakata, T., & Trehub, S. E. (2004). Infants' responsiveness to maternal speech and singing. *Infant Behavior & Development, 27*(4), 455–464.
23. Corbeil, M., Trehub, S. E., & Peretz, I. (2016). Singing delays the onset of infant distress. *Infancy, 21*, 373–391. https://doi.org/10.1111/infa.12114
24. Zentner, M., & Eerola, T. (2010). Rhythmic engagement with music in infancy. *Proceedings of the National Academy of Sciences, 107*, 5768–5773.
25. Cirelli, L. K., Einarson, K. M., & Trainor, L. J. (2014). Interpersonal synchrony increases prosocial behavior in infants. *Developmental Science, 17*(6), 1003–1011. https://doi.org/10.1111/desc.12193. PMID: 25513669.
26. Lense, M. D., Shultz, S., Astésano, C., & Jones, W. (2022). Music of infant-directed singing entrains infants' social visual behavior. *Proceedings of the National Academy of Sciences, 119*(45), e2116967119.
27. Trehub, S. E., Bull, D., & Thorpe, L. A. (1984). Infants' perception of melodies: The role of melodic contour. *Child Development, 55*, 821830.

28. Trehub, S. E., Thorpe, L. A., & Morrongiello, B. A. (1985). Infants' perception of melodies: Changes in a single tone. *Infant Behavior and Development., 8*, 213–223. https://doi.org/10.1016/S0163-6383(85)80007-2
29. Trehub, S. E., Thorpe, L. A., & Trainor, L. J. (1990). Infants' perception of good and bad melodies. *Psychomusicology, 9*, 5–19.
30. Trehub, S. E., Schellenberg, E. G., & Kamenetsky, S. B. (1999). Infants' and adults' perception of scale structure. *Journal of Experimental Psychology: Human Perception and Performance, 25*, 965–975.
31. Chang, H. W., & Trehub, S. E. (1977a). Infants' perception of temporal grouping in auditory patterns. *Child Development, 48*, 16661670.
32. Hannon, E. E., & Trehub, S. E. (2005a). Tuning in to rhythms: Infants learn more readily than adults. *Proceedings of the National Academy of Sciences (USA), 102*, 12639–12643.
33. Saffran, J. R., Loman, M. M., & Robertson, R. R. W. (2000). Infant memory for musical experiences. *Cognition, 77*, 15.
34. Trainor, L. J., Wu, L., & Tsang, C. D. (2004). Long-term memory for music: Infants remember tempo and timbre. *Developmental Science, 7*, 289–296. 35 Volkova, A.
35. Trehub, S. E., & Schellenberg, E. G. (2006). Infants' memory for musical performances. *Developmental Science, 9*, 584–590.
36. Swanwick, K., & Tillman, J. B. (1986). The sequence of musical development: A study of children's composition. *British Journal of Music Education, 3*, 305–339.
37. Darwin, C. R. (1872). *The origin of species by means of natural selection, or the preservation of favoured races in the struggle for life*. John Murray. 6th edition; with additions and corrections. Eleventh thousand.
38. Gould, S. J., & Vrba, E. S. (1982). Exaptation—A missing term in the science of form. *Paleobiology, 8*, 4–15.
39. Prum, R. O. (1999). Development and evolutionary origin of feathers. *The Journal of Experimental Zoology, 285*, 291–306.
40. Trainor, L. (2015). The origins of music in auditory scene analysis and the roles of evolution and culture in musical creation. *Philosophical Transactions of the Royal Society of London. Series B, Biological Sciences, 370*. https://doi.org/10.1098/rstb.2014.0089
41. Pinker, S. (1997). *How the mind works*. W W Norton & Co.
42. Patel, A. D. (2018). Music as a transformative technology of the mind: An update. In H. Honing (Ed.), *The origins of musicality* (pp. 113–126). The MIT Press.

43. Shilton, D. (2022). *Sweet participation: The evolution of music as an interactive technology* (Vol. 5, p. 205920432210847). SAGE Publishing.
44. Laland, K., Matthews, B., & Feldman, M. W. (2016). An introduction to niche construction theory. *Evolutionary Ecology, 30*, 191–202. https://doi.org/10.1007/s10682-016-9821-z. Epub 2016 Feb 3. PMID: 27429507; PMCID: PMC4922671.
45. Levins, R., & Lewontin, R. C. (1985). *The dialectical biologist*. Harvard University Press.
46. Wrangham, R. W., Jones, J. H., Laden, G., Pilbeam, D., & Conklin-Brittain, N. (1999). The raw and the stolen: Cooking and the ecology of human origins. *Current Anthropology, 40*(5), 567–594.
47. Dunbar, R. I. (2017). Group size, vocal grooming and the origins of language. *Psychonomic Bulletin & Review, 24*(1), 209–212. https://doi.org/10.3758/s13423-016-1122-6. PMID: 27460463.
48. Dissanayake, E. (2009). Root, leaf, blossom, or bole: Concerning the origin and adaptive function of music. In S. Malloch & C. Trevarthen (Eds.), *Communicative musicality: Exploring the basis of human companionship* (pp. 17–30). Oxford University Press.
49. Dissanayake, E. (2000). Antecedents of the temporal arts in early mother–infant interaction. In N. L. Wallin, B. Merker, & S. Brown (Eds.), *The origins of music* (pp. 389–410). The MIT Press.
50. Bradshaw, G. A., & Schore, A. N. (2007). How elephants are opening doors: Developmental neuroethology, attachment and social context. *Ethology, 113*, 426–436.
51. Miller, G. (2000). Evolution of human music through sexual selection. In N. L. Wallin, B. Merker, & S. Brown (Eds.), *The origins of music* (pp. 329–360). The MIT Press.
52. Darwin, C. R. (1871). *The descent of man, and selection in relation to sex*. John Murray.
53. Mehr, S. A., Krasnow, M. M., Bryant, G. A., & Hagen, E. H. (2020). Origins of music in credible signaling. *The Behavioral and Brain Sciences, 44*, e60. https://doi.org/10.1017/S0140525X20000345. PMID: 32843107; PMCID: PMC7907251.
54. Savage, P. E., Loui, P., Tarr, B., Schachner, A., Glowacki, L., Mithen, S., & Fitch, W. T. (2021). Music as a coevolved system for social bonding. *Behavioral and Brain Sciences*, Cambridge Core, *44*, e59. https://doi.org/10.1017/S0140525X20000333

55. Schäfer, T., Sedlmeier, P., Städtler, C., & Huron, D. (2013). The psychological functions of music listening. *Frontiers in Psychology, 4*, 511. https://doi.org/10.3389/fpsyg.2013.00511. PMID: 23964257; PMCID: PMC3741536.
56. Henkjan, H., ten Cate, C., Isabelle, P., & Trehub Sandra, E. (2015). Without it no music: Cognition, biology and evolution of musicality. *Philosophical Transactions of the Royal Society B: Biological Sciences, 370*, 1664.
57. Brown, S. (1999). The "Musilanguage" model of music evolution. In N. L. Wallin, B. Merker, & S. Brown (Eds.), *The origins of music*. The MIT Press.

3

Evolving Concepts of Autism

Abstract In this chapter I describe changes in diagnostic classification, and research investigating biological and environmental factors that have been associated with an increased likelihood of autism. I describe theoretical and empirical work on early development in autism, and discuss the rapidly increasing importance of research carried out within the neurodiversity paradigm.

Keywords Diagnostic classification • Diagnostic and Statistical Manual • International Classification of Diseases • Neurotypical responses

Brief History of Autism

The emergence of autism as a concept is rooted in advances in adult psychiatry during the late eighteenth and early nineteenth centuries. In 1911, the Swiss psychiatrist Paul Eugen Bleuler (1857–1939) published *Dementia Praecox or the Group of Schizophrenias* [1] and described four symptoms characteristics of this group of psychiatric conditions. These included (1) a disturbance in the capacity for making normal associations, (2) difficulties in affectivity and flattened affect, (3) ambivalence in response to conflicting feelings or attitudes, and (4) autism, defined as a separation of the individual from her/his reality. During this period schizophrenia was diagnosed in children as well as adults, and in addition to the disturbances in thought and feeling described by Bleuler, diagnostic criteria for childhood schizophrenia included a retraction of interest from the environment and a tendency to perseveration or stereotypy [2]. In his first paper on autism Kanner [3] reported 'symptoms' of 'extreme aloneness, obsessiveness, stereotypy and echolalia' and considered how these might differ from the symptoms characteristic in childhood schizophrenia. He noted that his patients were 'from the start anxiously and tensely impervious to people', whilst for children with schizophrenia the onset of symptoms was 'preceded by at least two years of essentially average development with a more or less gradual change in behaviour'. According to Kanner, children with schizophrenia stepped 'out of a world' in which they had 'been a part', whilst the children he described cautiously extended feelers into a world in which they had always been strangers. This was reflected in changes in behaviour. For example, at 5 to 6 years, echolalia, the repetition of another person's spoken words, was replaced with a more communicative form of interaction, and at around 6–8 years the children engaged in play, albeit on the periphery alongside groups of other children. Kanner also described how 'monotonously repetitious' behaviours and verbal utterances were gradually replaced by 'obsessive preoccupations'. Kanner's 1943 study provided rich descriptive data, and in 1956, in collaboration with Leon Eisenberg, he outlined diagnostic features and essential criteria for autism [4]. Essential diagnostic features included a 'profound lack of affective contact with others', an 'obsessive desire for sameness', a fascination with objects, language 'impairment' or language used without communicative intent, and atypical cognitive skills. The defining and essential diagnostic criteria were a 'profound lack of affective contact and repetitive, ritualistic behavior'.

Working during the same time period as Kanner, Hans Asperger (1906–1980), an Austrian paediatrician, had begun to study a group of verbally able male children, who displayed what he described as 'a lack of empathy, little ability to form friendships, one-sided conversations, intense absorption in a special interest and clumsy movements'. Although Asperger had begun to disseminate his ideas about these children as early as 1938 [5] and had published his findings in 1944 [6], he lectured and wrote in German, during the Second World War period, and his work was initially less widely disseminated than that of Kanner and his colleagues. However, the significance of Asperger's work in relation to the work carried out by Kanner was noted by some clinicians working in the field. For example, in 1962, Van Krevelen and Kuipers [7] discussed both accounts and concluded that whilst Kanner's condition was a 'psychotic process' with a time course, Asperger had described static personality traits, characterised by rationality and reduced sensitivity. In a later article Van Krevelen (1971) [8] distinguished between what he described as 'early infantile autism' and 'autistic psychopathy', citing differences in symptom onset, eye contact, motor and language skills, social interaction skills and social prognosis. Following the publication of Lorna Wing's 1981 [9] paper describing Asperger's findings, and her observations of children whose cognitive strengths and behavioural profiles more closely resembled those described by Asperger than those described by Kanner, the term Asperger's syndrome came into use, and was included in the fourth edition of the *Diagnostic and Statistical Manual of Mental Disorders* (DSM-1V, 1994) [10].

Changes in the Diagnostic Classification of Autism

The *Diagnostic and Statistical Manual* (DSM), published by the American Psychiatric Association (APA), and the International Classification of Diseases (ICD), published by the World Health Organisation, are used worldwide with the aim of increasing consistency across clinical and research settings. DSM-II, published in 1968 [11], included the category 'schizophrenia, childhood type' and DSM-III (1980) [12] included the category 'infantile autism' as a condition that was distinct from childhood schizophrenia. In 1987 DSM-III was revised (DSM-III-R (APA) 1987) and the term 'infantile autism' was replaced with 'autistic disorder'.

This change in terminology reflected new insights gained from earlier and ongoing research focused on identifying difficulties and strengths in children with special educational needs [13]. The criteria for 'infantile autism' included 16 items grouped in three categories of behaviours. These were (1) impaired social development that included characteristics that were not commensurate with the child's intellectual level, (2) delayed and deviant language development that had specific features and was not commensurate with the child's intellectual level, and (3) an 'insistence on sameness' defined as stereotyped play patterns, abnormal preoccupations or resistance to change. A fourth criterion specified that these difficulties must be in evidence before the child reached 30 months.

Both DSM and ICD manuals are periodically reviewed and revised with the aim of ensuring that criteria reflect scientific advances, are valid and reliable, and have clinical utility. The publication of both DSM-V (2013) [14] and ICD-11 (2019) [15] marked a radical shift away from categorical descriptions of autism and related conditions. DSM-IV (Text Revision, 2000) [16] had included (a) autistic disorder (AD), (b) Asperger's syndrome (AS), (c) Rett's disorder, (d) Childhood disintegrative disorder and (e) Pervasive Developmental Disorder Not Otherwise Specified (PDD-NOS), all grouped under the heading Pervasive Developmental Disorders (PDD). However, research had failed to validate these subcategories [17]. For example, Rett's disorder was now understood to be a progressive neurological condition with a distinct genetic etiology, and PDD-NOS, a separate subcategory for those who didn't meet criteria for AD, had been found to lack utility from clinical and research perspectives. Research had largely focused on investigating AD and AS, and little had been learned about the characteristics and needs of the substantial group of individuals diagnosed with PDD-NOS. The removal of the AS subcategory in DSM-5 resulted from research showing that individuals with this diagnosis could not be reliably distinguished from those with AD who did not have cognitive impairment. Indeed, Lord and colleagues [17] reported that the factor most strongly predicting whether an individual was diagnosed with AS, or what was then termed high-functioning autism, was the clinic where diagnosis had taken place.

In addition to subsuming PDD subcategories under the umbrella term autism spectrum disorder (ASD), DSM-V introduced changes in criteria and the way difficulties were documented by clinicians. Moving away from the traditional 'triad of impairments' in which communication and socialisation difficulties were grouped independently, diagnostic criteria were now listed under two headings: (1) socio-communicative impairments and (2) restricted and repetitive behaviours, interests and activities. The second group of criteria were also expanded to include 'Hyper- or hypo-reactivity to sensory input or unusual interests in sensory aspects of the environment'. Developmental theories of autism, outlined later in this chapter, raise questions about the validity of conceptualising sensory disturbance as a discrete behavioural characteristic within the 'restricted and repetitive behaviours' grouping. According to these theories, early sensory and sensorimotor difficulties influence all aspects of development including what have been conceptualised as 'core social-communication impairments'. DSM-V did not include intellectual and language 'impairments' in core criteria, though it did include specifiers that required clinicians to document these and other co-occurring conditions. Clinicians were also required to specify each individual's support needs ranging between level one (support is needed) and level three (very substantial support is needed). Finally, it was specified that the presence or absence of catatonia, known medical or genetic conditions, or environmental factors should be recorded. Physiological conditions that are commonly reported in individuals who meet criteria for ASD include sleep disorders [18], gastrointestinal disorders [19] and epilepsy [20]. Other conditions that commonly co-occur with autism include attention deficit hyperactivity disorder (ADHD) [21], obsessive compulsive disorder [22] and different categories of anxiety conditions [23, 24]. Difficulties with executive functions, including attentional control, inhibitory control, working memory and cognitive flexibility, are also widely reported in autism, although the severity and profiles of these difficulties show considerable differences across individuals [25].

It is currently estimated that autism is around three times more prevalent among males than among females [26]. The female protective effect, in which females are required to carry a greater genetic and/or environmental load than males to meet diagnostic criteria, has been proposed as an explanation for this gender ratio difference [27]. However, the reliability of past research on autistic females is brought into question by a well-documented diagnostic gender bias [28] that has resulted in an under-representation of females in research studies [29]. Gender diversity is significantly increased in autistic compared with neurotypical populations [30–32], and research has shown that gender as well as sex influence autism presentation and the diagnostic process [33, 34]. Understanding how autism presentation is influenced by sex and gender is an important and rapidly evolving field of enquiry within autism science [35–38].

The publication of DSM-II (1968) marked the first inclusion of the condition now described as 'autism spectrum disorder' under the heading of 'schizophrenia, childhood type'. This period also witnessed the emergence of the autism advocacy movement. Initially driven by parents advocating for the provision of appropriate educational and therapeutic services for their children, the later availability of the Internet enabled the emergence and proliferation of online autistic communities that became increasingly important for the autism self-advocacy movement [39, 40]. During the 1990s Robin Singer, a sociologist and autism activist, coined the term neurodiversity and her writings motivated an ongoing critical appraisal of the way neurological differences had been conceptualised in the traditional biomedical model [41]. In 1993, Jim Sinclair published 'Don't Mourn for Us' [42] and challenged the way autism was conceptualised as a source of grief for the parents of autistic children [43]. Singer's work laid the foundations for evolving narratives about how differences in ways of experiencing and engaging with the world are conceptualised [44]. The motto 'Nothing about us, without us' became an important rallying call in disability activism during the 1990s, and autistic researchers working within the neurodiversity paradigm began to play an increasingly central role in shaping the goals and methods used within research, therapeutic and educational settings [45, 46]. A further important marker of this societal change was the participation of Ari Ne'eman and Dr Steven Kapp, representatives of the Autistic Self Advocacy Network, in the DSM-V revision working group [47].

Approaches to Understanding Autism

Early conceptualisations of autism were formulated during the period when psychiatry was dominated by psychoanalysis, and this inevitably shaped early thinking about its causes. The first serious challenge to psychodynamic explanations for autism was mounted by Bernard Rimland, an experimental psychologist and parent of a child with autism. In his book *Infantile Autism* (1964) [48] he challenged the assumption that biology played no causal role in autism and highlighted the negative impact that scientifically unsubstantiated narratives around 'parent-blaming' had on autistic children and their family members. In 1977, Folstein and Rutter [49] provided the first scientific evidence for a substantial genetic influence on autism. The study investigated autism prevalence rates in groups of identical (monozygotic) and fraternal (dizygotic) twin pairs and reported that four of the 11 pairs of identical twins both had autism whilst autism concordance was not observed in any of the 10 fraternal twin pairs. Folstein and Rutter's original finding has been replicated in multiple large-scale twin studies [50]. For example, in a study of 336 identical twin pairs, Castlebaum, Sylvester, Zhang and Yu (2020) [51] reported a very high autism heritability rate. However, consistent with the findings from earlier twin studies [52], large within-pair differences on measures of social communication and interaction were also identified and suggested that the factors responsible for heritability are different to those responsible for symptom severity. Lichtenstein, Carlström, Råstam, Gillberg and Anckarsäter (2010) [53] studied a sample (n = 10,895) of Swedish twins born between 1992 and 2000 and found that identical twins showed high concordance rates not only for ASD, but also for attention deficit hyperactivity disorder (ADHD), developmental coordination disorder and tic disorders. The data analysis showed that the probability of identical autistic co-twins both having a diagnosis of ADHD was 44%, compared with 15% for fraternal co-twins. Differences in cross-disorder effects between identical and fraternal twins were observed for most other co-conditions, and substantial proportions of the genetic variance for ASD were shared with each of the other conditions. The authors concluded that their results suggest a general genetic susceptibility for developmental difficulties, rather than a specific genetic susceptibility for autism.

Research in genetics has associated a large number of genes with an increased likelihood of autism [54, 55]. However, none of these genes are consistently mutated in every person with autism, none link with autism every time they are mutated, and many are associated with an increased likelihood of other conditions [56–60]. Maternal health conditions, for example obesity, diabetes, pre-eclampsia and asthma, are also linked with an increased likelihood of autism [61]. These conditions may result in a decreased oxygen supply to the fetus during gestation or delivery, with consequences for the formation and development of the brain and changes in the expression and processing of genes involved in brain function. An important finding to emerge in genetics studies is that many of the genes that have been associated with autism are expressed in the cortex at very early stages of development. According to Courchesne, Pramparo, Gazestani and colleagues (2019) [62], approximately 94% of these genes are expressed during prenatal life, first during the earliest stages when cell proliferation, differentiation, migration and organisation occurs, and secondly, during peri- and post-natal stages when synapses and neural pathways are generated.

Research Into Early Development in Autism

During the last three decades the prospective longitudinal study design has been widely used to track early development in infants who have an autistic sibling and an increased likelihood (around 20%) of a later autism diagnosis [63]. In these studies infants are assessed at different time points over the first three years of life and are compared, on different developmental measures, with infants who do not have an autistic sibling. Evidence from prospective longitudinal studies, together with findings from clinical case reports and analyses of home videotapes, has revealed disruptions in patterns of development in infants later diagnosed with autism. In a recent review of this work Dawson, Rieder and Johnson (2023) [64] described four domains within which atypical development has been identified. First were difficulties with attentional flexibility.

These are typically manifested in a reduced propensity to sustain an external locus of attention and to flexibly switch attention between tasks, objects and/or people [65–67]. Second were alterations in patterns of social attention and engagement. Infants subsequently diagnosed with autism showed a gradual reduction in attention to faces, eyes and voices over the first year of life [68, 69]. Third were differences in prelinguistic development, including early vocalisations, gestures and joint attention. Subsequently diagnosed autistic infants showed delays in the onset of speech-like (canonical) vocalisations at 9–12 months [70] and used fewer gestures and gesture–speech combinations than typically developing infants [71]. The fourth category of developmental difference was in sensory and motor skills. Sensory differences (hyper- and hyposensitivity) in response to visual, auditory and tactile stimuli were observed by 6 months [72], and delays in sitting, pull to sit, reach to grasp and goal-directed reaching were also reported in one study [73].

Shultz, Klin and Jones (2015) [74] have contextualised what has been learned from prospective longitudinal studies in the context of developmental transitions that culminate in complex and contingent social interactions between typically developing infants and their carers. From hours after birth, infants display a range of reflexive behaviours that maximise the attention of carers and rapidly transition into more complex forms. For example, reflexive face tracking exhibited shortly after birth declines at around 4–6 weeks, with volitional attention to faces emerging at around two months. Caregivers' responses to their infants also increase in complexity as transitions from reflexive responses to volitional acts occur. For example, parental imitation of the infants' facial and vocal expressions elicits attention in infants and provides them with rich learning opportunities. According to this account, typically developing infants' transitions from reflexive behaviors to contingent volitional actions are established in mutually adaptive interactions between infants and carers. Although the first signs of sensorimotor and other disturbances in autism are not typically observed until later in the first year, development is an iterative process, and Schulz and colleagues argue that the effects of small and early disruptions in adaptive joint behaviours are likely to become larger/more observable over time. They also propose that such disruptions may be reflected in a co-opting of cortical mechanisms of attention to other non-social aspects of the environment.

A second theoretical model, developed by Johnson, Jones and Gliga (2015) [75] and Johnson (2017) [76], also builds on research reporting alterations in patterns of early sensory and sensorimotor skills in infants later diagnosed with autism. According to this account, genetic and environmental factors, singly or in tandem, cause widespread disruption in neural processing during the first year of life in autism. This disruption is manifested in 'neural noise' or patterns of neural firing that are not directly linked with a specific stimulus or activity. This alters the way the infant perceives and experiences its early environment and influences the trajectory of brain development and specialisation. Within this model, differences in perceptual, social and non-social skills in autism reflect the co-action of early disruptions in neural processing and later-occurring adaptive and compensatory responses and processes. An important finding to emerge in longitudinal studies is that patterns of development may show considerable individual differences in infants later diagnosed with autism, and this has significant implications for the design of early interventions [77, 78]. In a recent paper by Johnson, Charman, Pickles and Jones (2021) [79], the authors outline an innovative methodological framework within which early disruptions that influence the infants' experiences within its environment may be distinguished from later developing mechanisms (neurocognitive modifiers) that potentially serve compensatory functions. The first of these is executive attention. This mechanism has an increasing influence on brain development towards the end of the first year and may enable the infant to re-orient attention away from distracting information. The second neurocognitive modifier is the brain system that promotes attention to information with social content. According to this account, strong skills within this domain may promote attention to social stimuli and ensure the levels of exposure necessary for the establishment of specialised brain regions and networks dedicated to processing social information. These theoretical accounts situate evidence from prospective longitudinal studies in the context of considerable recent advances in the field of developmental cognitive neuroscience.

Changing Perspectives

An important perspective change that reflects the activities of autistic scholars, and of scientists, many of them autistic, working within the neurodiversity paradigm has been to acknowledge the powerful influence that cultural beliefs have on science. Writing from the perspective of clinicians focused on early development in autism, Klin, Micheletti, Kaliman, Schultz, Constantino and Jones (2020) [80] have described how the implementation of services such as early interventions may be constrained by culturally entrenched but scientifically invalidated beliefs about the nature and 'causes' of autism. Within the biomedical model, autism was conceptualised as a rare 'disease' 'caused' by gene-brain pathology that influences outcome in a deterministic way. However, autism is now recognised as a relatively common condition, and research findings implicate complex interactions between genetic, epigenetic, environmental and developmental factors in its emergence. Lord and colleagues (2021) [81] have estimated that autism affects around 78 million people worldwide, and they stress the importance of implementing changes that will promote positive quality of life outcomes for autistic people. They write that 'such change will depend on investments in science focused on practical clinical issues, and on social and service systems that acknowledge the potential for change and growth as well as the varied, complex needs of the autistic individuals and their families whose lives could be changed with such an effort'.

In contrast to the biomedical model that situates disability within individuals, the social model focuses on identifying and eradicating barriers that constrain an individual's active and positive participation in everyday life. In addition to the commitments embodied in the social model of disability, the neurodiversity movement acknowledges neural diversity as a natural and valuable form of human variation [82]. Writing on the neurodiversity movement and autism, den Houting (2019) [83] has described how 'disability results not from autism itself but instead from living in a society which tends to be physically, socially and emotionally inhospitable towards autistic people'. Research investigating the social and emotional barriers that den Houting describes has highlighted the powerful and negative effects that stereotyping and prejudice have on the wellbeing of autistic people [84]. In a study carried out by Botha, Dibb and Frost (2020) [85] autistic adults described their experiences of stigma in everyday life. They described feelings of tension that resulted from a discrepancy between their own value-neutral perception of their autism and the negative stereotypes of autism held in the wider society. They reported how they attempted to manage this tension using disclosure and masking strategies. Neurotypical responses to an autistic person's disclosure may be positive or negative, but anticipation of negative reactions and evaluations remain a potential source of anxiety. However, despite the risks associated with disclosure, autistic participants in several studies have described the importance of disclosure as a means of raising awareness and reframing social perceptions of autism within the wider society [86, 87]. Masking or camouflaging may be motivated by a desire for social acceptance and autistic people who engage in masking may present as behaviorally non-autistic. However, this strategy incurs considerable personal costs including mental exhaustion and psychological distress [88, 89]. Indeed evidence from a survey study carried out by Cassidy, Bradley, Shaw and Baron-Cohen (2018) [90] showed that masking autistic symptoms was a significant predictor of suicidality amongst autistic people. Consistent with predictions from the social model, these problems are not an inevitable consequence of autism, but result from societal barriers to acceptance and inclusion. A basic principle of the neurodiversity movement is that disability can be minimised or avoided when the barriers that negatively impact on everyday life are removed. Studies designed by or carried out in collaboration with autistic people are increasingly identifying ways that physical modifications within home, school and workplace settings can improve learning, productivity and wellbeing [91, 92]. Findings linking psychological disturbance and distress with outdated and stereotyped concepts of autism highlight the importance of social change for ensuring better outcomes for autistic people.

References

1. Bleuler, E. (1950). *Dementia praecox or the group of schizophrenias*. International Universities Press.
2. Potter, H. W. (1933). Schizophrenia in children. *American Journal of Psychiatry, 89*, 1253–1270.
3. Kanner, L. (1943). Autistic disturbances of affective contact. *Nervous Child, 2*, 217–250.
4. Eisenberg, L., & Kanner, L. (1956). Childhood schizophrenia: Symposium, 1955: 6. Early infantile autism, 1943–55. *American Journal of Orthopsychiatry, 26*(3), 556–566. https://doi.org/10.1111/j.1939-0025.1956.tb06202.x
5. Silberman, S. (2015). Neurotribes: The legacy of autism and the future of neurodiversity. *Anthropological Quarterly, 88*, 1111–1121. https://doi.org/10.1353/anq.2015.0057
6. Asperger, H. (1944). Die "AutistischenPsychopathen" imKindesalter. [The "autistic psychopaths" in childhood]. *Archiv für Psychiatrie und Nervenkrankheiten, 117*, 76–136. https://doi.org/10.1007/BF01837709
7. Van Krevelen, D. A., & Kuipers, C. (1962). The psychopathology of autistic psychopathy. *Acta Paedopsychiatrica: International Journal of Child & Adolescent Psychiatry, 29*(1), 22–31.
8. Van Krevelen, D. A. (1971). Early infantile autism and autistic psychopathy. *Journal of Autism and Childhood Schizophrenia, 1*, 82–86.
9. Wing, L. (1981). Asperger's syndrome: A clinical account. *Psychological Medicine, 11*(1), 115–129. https://doi.org/10.1017/s0033291700053332. PMID: 720873.
10. American Psychiatric Association (Ed.). (1994). *Diagnostic and statistical manual of mental disorders* (4th ed.). American Psychiatric Association Press.
11. American Psychiatric Association (Ed.). (1968). *Diagnostic and statistical manual of mental disorders* (2nd ed.). American Psychiatric Association.
12. American Psychiatric Association (Ed.). (1980). *Diagnostic and statistical manual of mental disorders* (3rd ed.). American Psychiatric Association.
13. Wing, L., & Gould, J. (1979). Severe impairments of social interaction and associated abnormalities in children: Epidemiology and classification. *Journal of Autism and Developmental Disorders, 9*(1), 11–29. https://doi.org/10.1007/BF01531288
14. American Psychiatric Association (Ed.). (2013). *Diagnostic and statistical manual for mental disorders* (5th ed.). American Psychiatric Association.

15. World Health Organization. (2019). *International statistical classification of diseases and related health problems* (11th ed.). https://icd.who.int). World Health Organization.
16. American Psychiatric Association (Ed.). (2000). *Diagnostic and statistical manual of mental disorders* (4th ed.). American Psychiatric Association. Text Revision.
17. Lord, C., Petkova, E., Hus, V., Gan, W., Lu, F., Martin, D. M., Ousley, O., Guy, L., Bernier, R., Gerdts, J., Algermissen, M., Whitaker, A., Sutcliffe, J. S., Warren, Z., Klin, A., Saulnier, C., Hanson, E., Hundley, R., Piggot, J., Fombonne, E., Steiman, M., Miles, J., Kanne, S. M., Goin-Kochel, R. P., Peters, S. U., Cook, E. H., Guter, S., Tjernagel, J., Green-Snyder, L. A., Bishop, S., Esler, A., Gotham, K., Luyster, R., Miller, F., Olson, J., Richler, J., & Risi, S. (2012). A multisite study of the clinical diagnosis of different autism spectrum disorders. *Archives of General Psychiatry, 69*(3), 306–313. https://doi.org/10.1001/archgenpsychiatry.2011.148. Epub 2011 Nov 7. PMID: 22065253; PMCID: PMC3626112.
18. Mazurek, M. O., et al. (2019). Course and predictors of sleep and co-occurring problems in children with autism spectrum disorder. *Journal of Autism and Developmental Disorders, 49*, 2101–2115.
19. Kang, V., Wagner, G. C., & Ming, X. (2014). Gastrointestinal dysfunction in children with autism spectrum disorders. *Autism Research, 7*(4), 501–506.
20. Tuchman, R., & Rapin, I. (2002). Epilepsy in autism. *The Lancet Neurology, 1*(6), 352–358, ISSN 1474-4422. https://doi.org/10.1016/S1474-4422(02)00160-6
21. Rong, Y., Yang, C.-J., Jin, Y., & Wang, Y. (2021). Prevalence of attention-deficit/hyperactivity disorder in individuals with autism spectrum disorder: A meta-analysis. *Research in Autism Spectrum Disorder, 83*, 101759. https://doi.org/10.1016/j.rasd.2021.101759
22. Meier, S. M., Petersen, L., Schendel, D. E., Mattheisen, M., Mortensen, P. B., & Mors, O. (2015). Obsessive-compulsive disorder and autism spectrum disorders: Longitudinal and offspring risk. *PLoS One, 10*(11), e0141703. https://doi.org/10.1371/journal.pone.0141703. PMID: 26558765; PMCID: PMC4641696.
23. Kerns, C. M., Winder-Patel, B., Iosif, A. M., Nordahl, C. W., Heath, B., Solomon, M., & Amaral, D. G. (2021). Clinically significant anxiety in children with autism spectrum disorder and varied intellectual functioning. *Journal of Clinical Child & Adolescent Psychology, 50*(6), 780–795.

24. Hollocks, M., Lerh, J., Magiati, I., Meiser-Stedman, R., & Brugha, T. (2019). Anxiety and depression in adults with autism spectrum disorder: A systematic review and meta-analysis. *Psychological Medicine, 49*(4), 559–572. https://doi.org/10.1017/S0033291718002283
25. Demetriou, E. A., DeMayo, M. M., & Guastella, A. J. (2019). Executive function in autism spectrum disorder: History, theoretical models, empirical findings, and potential as an endophenotype. *Frontiers in Psychiatry, 10*, 753. https://doi.org/10.3389/fpsyt.2019.00753. PMID: 31780959; PMCID: PMC6859507.
26. Loomes, R., Hull, L., & Mandy, W. P. L. (2017). What is the male-to-female ratio in autism spectrum disorder? A systematic review and meta-analysis. *Journal of the American Academy of Child & Adolescent Psychiatry, 56*, 466. https://doi.org/10.1016/j.jaac.2017.03.013
27. Jacquemont, S., Coe, B. P., Hersch, M., et al. (2014). A higher mutational burden in females supports a "female protective model" in neurodevelopmental disorders. *American Journal of Human Genetics, 94*, 415–425.
28. Kirkovski, M., Enticott, P. G., & Fitzgerald, P. B. (2013). A review of the role of female gender in autism spectrum disorders. *Journal of Autism and Developmental Disorders, 43*(11), 2584–2603. https://doi.org/10.1007/s10803-013-1811-1
29. D'Mello, A. M., Frosch, I. R., Li, C. E., Cardinaux, A. L., & Gabrieli, J. D. E. (2022). Exclusion of females in autism research: Empirical evidence for a "leaky" recruitment-to-research pipeline. *Autism Research, 15*(10), 1929–1940. https://doi.org/10.1002/aur.2795. Epub 2022 Aug 22. PMID: 36054081; PMCID: PMC9804357.
30. Van Der Miesen, A. I., Hurley, H., & De Vries, A. L. (2016). Gender dysphoria and autism spectrum disorder: A narrative review. *International Review of Psychiatry, 28*(1), 70–80. https://doi.org/10.3109/09540261.2015.1111199. Epub 2016 Jan 12. PMID: 26753812.
31. Walsh, R. J., Krabbendam, L., Dewinter, J., & Begeer, S. (2018). Brief report: Gender identity differences in autistic adults: Associations with perceptual and socio-cognitive profiles. *Journal of Autism and Developmental Disorders, 48*(12), 4070–4078. https://doi.org/10.1007/s10803-018-3702-y. PMID: 30062396.
32. Nabbijohn, A. N., van der Miesen, A. I. R., Santarossa, A., Peragine, D., de Vries, A. L. C., Popma, A., Lai, M. C., & VanderLaan, D. P. (2019). Gender variance and the autism spectrum: An examination of children ages 6-12

years. *Journal of Autism and Developmental Disorders, 49*(4), 1570–1585. https://doi.org/10.1007/s10803-018-3843-z. PMID: 30547258.
33. McQuaid, G. A., Lee, N. R., & Wallace, G. L. (2022). Camouflaging in autism spectrum disorder: Examining the roles of sex, gender identity, and diagnostic timing. *Autism, 26*(2), 552–559. https://doi.org/10.1177/13623613211042131. Epub 2021 Aug 23. PMID: 34420418.
34. Lai, M. C., & Szatmari, P. (2020). Sex and gender impacts on the behavioural presentation and recognition of autism. *Current Opinion in Psychiatry, 33*(2), 117–123. https://doi.org/10.1097/YCO.0000000000000575. PMID: 31815760.
35. Hull, L., Petrides, K. V., & Mandy, W. (2020). The female autism phenotype and camouflaging: A narrative review. *Review Journal of Autism and Developmental Disorders, 7*, 306–317. https://doi.org/10.1007/s40489-020-00197-9
36. Calderoni, S. (2023). Sex/gender differences in children with autism spectrum disorder: A brief overview on epidemiology, symptom profile, and neuroanatomy. *Journal of Neuroscience Research, 101*(5), 739–750.
37. Wood-Downie, H., Wong, B., Kovshoff, H., Cortese, S., & Hadwin, J. A. (2021). Research review: A systematic review and meta-analysis of sex/gender differences in social interaction and communication in autistic and nonautistic children and adolescents. *Journal of Child Psychology and Psychiatry, 62*(8), 922–936.
38. Wood-Downie, H., Wong, B., Kovshoff, H., Mandy, W., Hull, L., & Hadwin, J. A. (2021). Sex/gender differences in camouflaging in children and adolescents with autism. *Journal of Autism and Developmental Disorders, 51*, 1353–1364.
39. Leadbitter, K., Buckle, K. L., Ellis, C., & Dekker, M. (2021). Autistic self-advocacy and the neurodiversity movement: Implications for autism early intervention research and practice. *Frontiers in Psychology, 12*, 635690. https://doi.org/10.3389/fpsyg.2021.635690. PMID: 33912110; PMCID: PMC8075160.
40. Dekker, M. (2020). From exclusion to acceptance: Independent living on the autistic spectrum. In S. Kapp (Ed.), *Autistic community and the neurodiversity movement: Stories from the frontline* (pp. 41–49). Springer Nature. https://doi.org/10.1007/978-981-13-8437-0_3
41. Singer, J. (1999). 'Why can't you be normal for once in your life?' From a 'problem with no name' to the emergence of a new category of difference. In

M. Corker & S. French (Eds.), *Disability discourse* (pp. 59–67). Open University Press.

42. Sinclair J (1993). Don't mourn for us. http://www.autreat.com/dont_mourn.html

43. Pripas-Kapit, S. (2020). Historicizing Jim Sinclair's "Don't mourn for us": A cultural and intellectual history of neurodiversity's first manifesto. In S. K. Kapp (Ed.), *Autistic community and the neurodiversity movement: Stories from the frontline* (pp. 23–39). Palgrave Macmillan.

44. Dwyer, P. (2022). The neurodiversity approach(es): What are they and what do they mean for researchers? *Human Development, 66*(2), 73–92. https://doi.org/10.1159/000523723. Epub 2022 Feb 22. PMID: 36158596; PMCID: PMC9261839.

45. Pellicano, E., & den Houting, J. (2022). Annual research review: Shifting from 'normal science' to neurodiversity in autism science. *Journal of Child Psychology and Psychiatry, 63*, 381–396. https://doi.org/10.1111/jcpp.13534

46. Kenny, L., Hattersley, C. M., Molins, B., Buckley, C., Povey, C., & Pellicano, E. (2016). Which terms should be used to describe autism? Perspectives from the UK autism community. *Autism, 20*, 442–462.

47. Kapp, S. K., & Ne'eman, A. (2020). Lobbying Autism's diagnostic revision in the DSM-5. In S. Kapp (Ed.), *Autistic community and the neurodiversity movement.* Palgrave Macmillan. https://doi.org/10.1007/978-981-13-8437-0_13

48. Rimland, B. (1964). *Infantile autism: The syndrome and its implications for a neural theory of behavior* (p. 61). Prentice-Hall.

49. Folstein, S., & Rutter, M. (1977). Infantile autism: A genetic study of 21 twin pairs. *Journal of Child Psychology and Psychiatry, 18*(4), 297–321. https://doi.org/10.1111/j.1469-7610.1977.tb00443.x. PMID: 562353.

50. Tick, B., Bolton, P., Happé, F., Rutter, M., & Rijsdijk, F. (2016). Heritability of autism spectrum disorders: a meta-analysis of twin studies. *Journal of Child Psychology and Psychiatry, 57*(5), 585–595. https://doi.org/10.1111/jcpp.12499. Epub 2015 Dec 27. PMID: 26709141; PMCID: PMC4996332.

51. Castelbaum, L., Sylvester, C. M., Zhang, Y., Yu, Q., & Constantino, J. N. (2020). On the nature of monozygotic twin concordance and discordance for autistic trait severity: A quantitative analysis. *Behavior Genetics, 50*(4), 263–272. https://doi.org/10.1007/s10519-019-09987-2. Epub 2019 Dec 18. PMID: 31853901; PMCID: PMC7355281.

52. Le Couteur, A., Bailey, A., Goode, S., Pickles, A., Robertson, S., Gottesman, I., & Rutter, M. (1996). A broader phenotype of autism: The clinical

spectrum in twins. *Journal of Child Psychology and Psychiatry, 37*(7), 785–801. https://doi.org/10.1111/j.1469-7610.1996.tb01475.x. PMID: 8923222.
53. Lichtenstein, P., Carlström, E., Råstam, M., Gillberg, C., & Anckarsäter, H. (2010). The genetics of autism spectrum disorders and related neuropsychiatric disorders in childhood. *The American Journal of Psychiatry, 167*(11), 1357–1363. https://doi.org/10.1176/appi.ajp.2010.10020223. Epub 2010 Aug 4. PMID: 20686188.
54. Satterstrom, et al. (2020). *Cell, 180*, 568–584.
55. Zhou, X., Feliciano, P., Shu, C., Wang, T., Astrovskaya, I., Hall, J. B., Obiajulu, J. U., Wright, J. R., Murali, S. C., Xu, S. X., Brueggeman, L., Thomas, T. R., Marchenko, O., Fleisch, C., Barns, S. D., Snyder, L. G., Han, B., Chang, T. S., Turner, T. N., Harvey, W. T., Nishida, A., O'Roak, B. J., Geschwind, D. H., SPARK Consortium, Michaelson, J. J., Volfovsky, N., Eichler, E. E., Shen, Y., & Chung, W. K. (2022). Integrating de novo and inherited variants in 42,607 autism cases identifies mutations in new moderate-risk genes. *Nature Genetics, 54*(9), 1305–1319. https://doi.org/10.1038/s41588-022-01148-2. Epub 2022 Aug 18. PMID: 35982159; PMCID: PMC9470534.
56. Gudmundsson, O. O., Walters, G. B., Ingason, A., Johansson, S., Zayats, T., Athanasiu, L., et al. (2019). Attention-deficit hyperactivity disorder shares copy number variant risk with schizophrenia and autism spectrum disorder. *Translational Psychiatry, 9*(1), 258. https://doi.org/10.1038/s41398-019-0599-y
57. Zarrei, M., Burton, C. L., Engchuan, W., Young, E. J., Higginbotham, E. J., MacDonald, J. R., et al. (2019). A large data resource of genomic copy number variation across neurodevelopmental disorders. *NPJ Genomic Medicine, 4*, 26. https://doi.org/10.1038/s41525-019-0098-3
58. Cross-Disorder Group of the Psychiatric Genomics Consortium. (2013). Identification of risk loci with shared effects on five major psychiatric disorders: A genome-wide analysis. *Lancet, 381*(9875), 1371–1379. https://doi.org/10.1016/S0140-6736(12)62129-1
59. Dias, C., & Walsh, C. A. (2020). Recent advances in understanding the genetic architecture of autism. *Annual Review of Genomics and Human Genetics, 21*, 289.
60. Grove, J., Ripke, S., Als, T. D., Mattheisen, M., Walters, R. K., Won, H., Pallesen, J., Agerbo, E., Andreassen, O. A., Anney, R., Awashti, S., Belliveau, R., Bettella, F., Buxbaum, J. D., Bybjerg-Grauholm, J., Bækvad-Hansen,

M., Cerrato, F., Chambert, K., Christensen, J. H., Churchhouse, C., Dellenvall, K., Demontis, D., De Rubeis, S., Devlin, B., Djurovic, S., Dumont, A. L., Goldstein, J. I., Hansen, C. S., Hauberg, M. E., Hollegaard, M. V., Hope, S., Howrigan, D. P., Huang, H., Hultman, C. M., Klei, L., Maller, J., Martin, J., Martin, A. R., Moran, J. L., Nyegaard, M., Nærland, T., Palmer, D. S., Palotie, A., Pedersen, C. B., Pedersen, M. G., dPoterba, T., Poulsen, J. B., Pourcain, B. S., Qvist, P., Rehnström, K., Reichenberg, A., Reichert, J., Robinson, E. B., Roeder, K., Roussos, P., Saemundsen, E., Sandin, S., Satterstrom, F. K., Davey Smith, G., Stefansson, H., Steinberg, S., Stevens, C. R., Sullivan, P. F., Turley, P., Walters, G. B., Xu, X., Autism Spectrum Disorder Working Group of the Psychiatric Genomics Consortium, BUPGEN, Major Depressive Disorder Working Group of the Psychiatric Genomics Consortium, 23andMe Research Team, Stefansson, K., Geschwind, D. H., Nordentoft, M., Hougaard, D. M., Werge, T., Mors, O., Mortensen, P. B., Neale, B. M., Daly, M. J., & Børglum, A. D. (2019). Identification of common genetic risk variants for autism spectrum disorder. *Nature Genetics, 51*(3), 431–444. https://doi.org/10.1038/s41588-019-0344-8. Epub 2019 Feb 25. PMID: 30804558; PMCID: PMC6454898.

61. Carter, S. A., Lin, J. C., Chow, T., Yu, X., Rahman, M. M., Martinez, M. P., Feldman, K., Eckel, S. P., Chen, J. C., Chen, Z., Levitt, P., Lurmann, F. W., McConnell, R., & Xiang, A. H. (2023). Maternal obesity, diabetes, preeclampsia, and asthma during pregnancy and likelihood of autism spectrum disorder with gastrointestinal disturbances in offspring. *Autism, 27*(4), 916–926. https://doi.org/10.1177/13623613221118430. Epub 2022 Sep 4. PMID: 36062479; PMCID: PMC9984567. (74).

62. Courchesne, E., Pramparo, T., Gazestani, V. H., et al. (2019). The ASD living biology: From cell proliferation to clinical phenotype. *Molecular Psychiatry, 24*, 88–107. https://doi.org/10.1038/s41380-018-0056-y

63. Rogers, S. J. (2009). What are infant siblings teaching us about autism in infancy? *Autism Research, 2*(3), 125–137. https://doi.org/10.1002/aur.81. PMID: 19582867; PMCID: PMC2791538.

64. Dawson, G., Rieder, A. D., & Johnson, M. H. (2023). A developmental social neuroscience perspective on infant autism interventions. *Annual Review of Developmental Psychology, 5*(1), 89–113.

65. Bradshaw, J., Klin, A., Evans, L., Klaiman, C., Saulnier, C., & McCracken, C. (2019). Development of attention from birth to 5 months in infants at

risk for autism spectrum disorder. *Development and Psychopathology, 32*(2), 1–11. https://doi.org/10.1017/S0954579419000233
66. Sacrey, L. A., Bryson, S. E., & Zwaigenbaum, L. (2013). Prospective examination of visual attention during play in infants at high-risk for autism spectrum disorder: A longitudinal study from 6 to 36 months of age. *Behavioural Brain Research, 256*, 441–450. https://doi.org/10.1016/j.bbr.2013.08.028
67. Elsabbagh, M., Fernandes, J., Webb, S. J., Dawson, G., Charman, T., Johnson, M. H., & British Autism Study of Infant Siblings, T. (2013). Disengagement of visual attention in infancy is associated with emerging autism in toddlerhood. *Biological Psychiatry, 74*(3), 189–194. https://doi.org/10.1016/j.biopsych.2012.11.030
68. Chawarska, K., Macari, S., & Shic, F. (2013). Decreased spontaneous attention to social scenes in 6-month-old infants later diagnosed with autism spectrum disorders. *Biological Psychiatry, 74*(3), 195–203. https://doi.org/10.1016/j.biopsych.2012.11.022
69. Jones, W., & Klin, A. (2013). Attention to eyes is present but in decline in 2–6-month-old infants later diagnosed with autism. *Nature, 504*(7480), 427–431.
70. Yankowitz, L. D., Petrulla, V., Plate, S., et al. (2022). Infants later diagnosed with autism have lower canonical babbling ratios in the first year of life. *Molecular Autism, 13*, 28. https://doi.org/10.1186/s13229-022-00503-8
71. Bradshaw, J., McCracken, C., Pileggi, M., Brane, N., Delehanty, A., Day, T., Federico, A., Klaiman, C., Saulnier, C., Klin, A., & Wetherby, A. (2021). Early social communication development in infants with autism spectrum disorder. *Child Development, 92*(6), 2224–2234. https://doi.org/10.1111/cdev.13683. PMID: 34786700; PMCID: PMC8935345.
72. Wolff, J. J., Dimian, A. F., Botteron, K. N., Dager, S. R., Elison, J. T., Estes, A. M., Hazlett, H. C., Schultz, R. T., Zwaigenbaum, L., Piven, J., & IBIS Network. (2019). A longitudinal study of parent-reported sensory responsiveness in toddlers at-risk for autism. *Journal of Child Psychology and Psychiatry, 60*(3), 314–324. https://doi.org/10.1111/jcpp.12978. Epub 2018 Oct 23. PMID: 30350375; PMCID: PMC8919956.
73. Dawson, G., Osterling, J., Meltzoff, A. N., & Kuhl, P. (2000). Case study of the development of an infant with autism from birth to two years of age. *Journal of Applied Developmental Psychology, 21*, 299.
74. Shultz, S., Klin, A., & Jones, W. (2018). Neonatal transitions in social behavior and their implications for autism. *Trends in Cognitive Sciences, 22*, 452. https://doi.org/10.1016/j.tics.2018.02.012
75. Johnson, M. H., Jones, E. J., & Gliga, T. (2015). Brain adaptation and alternative developmental trajectories. *Development and Psychopathology,*

27(2), 425–442. https://doi.org/10.1017/S0954579415000073. PMID: 25997763.
76. Johnson, M. H. (2017). Autism as an adaptive common variant pathway for human brain development. *Developmental Cognitive Neuroscience, 25*, 5–11. https://doi.org/10.1016/j.dcn.2017.02.004. Epub 2017 Feb 9. PMID: 28233663; PMCID: PMC6987822.
77. Ozonoff, S., Young, G., Landa, R., Brian, J., Bryson, S., Charman, T., Chawarska, K., Macari, S., Messinger, D., Stone, W., Zwaigenbaum, L., & Iosif, A.-M. (2015). Diagnostic stability in young children at risk for autism spectrum disorder: A baby siblings research consortium study. *Journal of Child Psychology and Psychiatry, and Allied Disciplines, 56*, 988. https://doi.org/10.1111/jcpp.12421
78. Ozonoff, S., Young, G. S., Brian, J., Charman, T., Shephard, E., Solish, A., & Zwaigenbaum, L. (2018). Diagnosis of autism spectrum disorder after age 5 in children evaluated longitudinally since infancy. *Journal of the American Academy of Child and Adolescent Psychiatry, 57*(11), 849–857.e2. https://doi.org/10.1016/j.jaac.2018.06.022. Epub 2018 Sep 3. PMID: 30392626; PMCID: PMC6235445.
79. Johnson, M. H., Charman, T., Pickles, A., & Jones, E. J. H. (2021). Annual research review: Anterior modifiers in the emergence of neurodevelopmental disorders (AMEND)-a systems neuroscience approach to common developmental disorders. *Journal of Child Psychology and Psychiatry, 62*(5), 610–630. https://doi.org/10.1111/jcpp.13372. Epub 2021 Jan 11. PMID: 33432656; PMCID: PMC8609429.
80. Klin, A., Micheletti, M., Klaiman, C., Shultz, S., Constantino, J. N., & Jones, W. (2020). Affording autism an early brain development re-definition. *Development and Psychopathology, 32*(4), 1175–1189. https://doi.org/10.1017/S0954579420000802. PMID: 32938507; PMCID: PMC7880583.
81. Lord, C., Charman, T., Havdahl, A., Carbone, P., Anagnostou, E., Boyd, B., Carr, T., de Vries, P. J., Dissanayake, C., Divan, G., Freitag, C. M., Gotelli, M. M., Kasari, C., Knapp, M., Mundy, P., Plank, A., Scahill, L., Servili, C., Shattuck, P., Simonoff, E., Singer, A. T., Slonims, V., Wang, P. P., Ysrraelit, M. C., Jellett, R., Pickles, A., Cusack, J., Howlin, P., Szatmari, P., Holbrook, A., Toolan, C., & McCauley, J. B. (2022). The Lancet Commission on the future of care and clinical research in autism. *Lancet, 399*(10321), 271–334. https://doi.org/10.1016/S0140-6736(21)01541-5. Epub 2021 Dec 6. Erratum in: Lancet. 2022 Dec 3; 400 (10367): 1926. PMID: 34883054.

82. Walker, N. (2012). Throw away the master's tools: Liberating ourselves from the pathology paradigm. In J. Bascom (Ed.), *Loud hands: Autistic people, speaking* (pp. 225–237). The Autistic Press.
83. den Houting, J. (2019). Neurodiversity: An insider's perspective. *Autism, 23*(2), 271–273. https://doi.org/10.1177/1362361318820762
84. Turnock, A., Langley, K., & Jones, C. R. G. (2022). Understanding stigma in autism: A narrative review and theoretical model. *Autism Adulthood., 4*(1), 76–91. https://doi.org/10.1089/aut.2021.0005. Epub 2022 Mar 9. PMID: 36605561; PMCID: PMC8992913.
85. Botha, M., Dibb, B., & Frost, D. M. (2022). "Autism is me": An investigation of how autistic individuals make sense of autism and stigma. *Disability & Society, 37*(3), 427. https://doi.org/10.1080/09687599.2020.1822782
86. Farsinejad, A., Russell, A., & Butler, C. (2022). Autism disclosure – The decisions autistic adults make. *Research in Autism Spectrum Disorders., 93*, 101936. https://doi.org/10.1016/j.rasd.2022.101936
87. Hull, L., Petrides, K. V., Allison, C., Smith, P., Baron-Cohen, S., Lai, M. C., & Mandy, W. (2017). "Putting on my best normal": Social camouflaging in adults with autism spectrum conditions. *Journal of Autism and Developmental Disorders, 47*(8), 2519–2534. https://doi.org/10.1007/s10803-017-3166-5. PMID: 28527095; PMCID: PMC5509825.
88. Miller, D., Rees, J., & Pearson, A. (2021). "Masking is life": Experiences of masking in autistic and nonautistic adults. *Autism in Adulthood, 3*, 330–338. https://doi.org/10.1089/aut.2020.0083
89. Cook, J., Hull, L., Crane, L., & Mandy, W. (2021, ISSN 0272-7358). Camouflaging in autism: A systematic review. *Clinical Psychology Review, 89*, 102080. https://doi.org/10.1016/j.cpr.2021.102080
90. Cassidy, S., Bradley, L., Shaw, R., & Baron-Cohen, S. (2018). Risk markers for suicidality in autistic adults. *Molecular Autism, 9*, 42. https://doi.org/10.1186/s13229-018-0226-4. PMID: 30083306; PMCID: PMC6069847.
91. Marcotte, J., Grandisson, M., & Milot, É. (2023). What makes home environments favorable to independence: Perspectives of autistic people and their parents. *Disability and Rehabilitation, 45*(10), 1684–1695. https://doi.org/10.1080/09638288.2022.2074153
92. Webster, A., & Roberts, J. (2022). Implementing the school-wide autism competency model to improve outcomes for students on the autism spectrum: A multiple case study of three schools. *International Journal of Inclusive Education, 26*(8), 796–814. https://doi.org/10.1080/13603116.2020.1735540

4

Music and Autism: Time for a Reappraisal?

Abstract Changing concepts of autism have prompted a re-evaluation of assumptions within theoretical models and research methodologies. Models of talent in autism do not reflect current thinking about the nature and development of musicality as a human trait. In this chapter first-person accounts of autistic people document rich intra- and interpersonal experiences of music. This provides a new starting point for understanding musical skills and talents in autism.

Keywords Contextualising musicality • Systematising model of autistic talents • Personal development • Creativity • Motivation

Contextualising Musicality in Autism

In Chap. 2 musicality was conceptualised as a fundamental human trait and explored in the context of Timbergen's questions about how it works, how it develops over time, why it evolved and the functions it currently serves. Researchers working in this area most frequently focus on addressing single questions, for example, about the nature of musicality in infancy [1] or the neural mechanisms associated with specific musical experiences [2]. This work can be accommodated within the broad biomusicology framework and has contributed much to our understanding of this complex human trait.

Within autism science, theoretical models do not differentiate between categories of skills and talents, and this has been reflected in methodological approaches to studying them. For example, in an early and extensive study exploring factors associated with talent and autism, Happé and Vital (2009) [3] asked the parents of six thousand 8-year-old twin pairs to report on their children's strengths within the domains of music, art and memory, and on cognitive and behavioural traits associated with autism. The results from this study showed that children with parent-reported skills in these domains displayed higher levels of restricted and repetitive behaviours and interests related to detail focus than children without these skills. The Weak Central Coherence [4], Veridical Mapping [5] and Systematising models [6], described in Chap. 1, all offer different accounts of the perceptual and cognitive mechanisms linked with autistic skills and talents. However, none gives consideration to the effect of social-emotional rewards that influence motivation levels of musical engagement and the developmental processes that shape musicality. The first evidence showing that autistic children are sensitive to musically expressed emotions was published a quarter of a century ago [7] and has been replicated in subsequent studies [8, 9]. In the most recent of these, Sivathasan, Dahary, Burack and Quintin (2023) [10] measured basic emotion recognition accuracy for faces, voices and music in autistic and typically developing (TD) children and reported that autistic children obtained higher identification scores than TD children in the music condition.

The systematising model of autistic talents describes strengths in extracting and representing structures within rule-based domains. However, research into musical development has shown that children's perception of form and structure develops over time and in tandem with their understanding of music's emotional connotations. For example, Kastner and Crowder (1990) [11] reported that by three years, typically developing children associate music written in the major mode with positive emotions (happy, contented) and music written in the minor mode with negative emotions (sad, angry). Strengths in understanding musically expressed emotions in autism may, in part be explained by early patterns of attention, and their impact on later emerging abilities. In a prospective longitudinal study Macari and colleagues (2021) [12] used eye tracking methods to measure attention to an experimenter's face in samples of 6-, 9- and 12-month-old infants with and without an elevated likelihood of autism. In the study infants viewed the examiner's face whilst s/he engaged with the infant using dynamic bids, songs, peek-a-boo, tickling and toy play. The results from the study showed that infants later diagnosed with autism attended to the examiner's face as intensely as infants without a later diagnosis in the song, toy play and peek-a-boo conditions. Although sensory, motor and other differences identified in studies of autistic infants may influence trajectories of musical development at later time points, early patterns of social attention and engagement, important for musical development in infancy, did not differ across autistic and TD groups in this study.

As described in chapter two, perception of music becomes increasingly complex over the lifespan. In Huron's elegant ITPRA model (2006) [13] of musical expectancies, he describes a sequence of imagination, tension, prediction, reaction and appraisal responses to unfolding temporal events during music listening. In a later neuroimaging study, Salimpoor, Zald, Zatorre, Dagher and McIntosh (2015) [14] demonstrated how these different music-driven responses are directly linked with patterns of activation in the reward centres of the brain. Music elicits emotions in multiple and diverse ways. Juslin and Västfjäll (2008) [15] and Juslin (2013) [16] have formulated theoretical models of the mechanisms that give rise to emotional responses to music. These include, for example, emotions that arise in response to music that elicits brain stem responses, rhythmic entrainment and visual imagery. Links between social events and music (e.g. the birthday party song) and personally meaningful events and music (autobiographical memory) give rise to emotions that can be experienced at different time points and in different contexts. Experiences of emotional contagion occur during group singing in religious, sporting and other communal settings. Research into the affective systems involved in musical perception and activity shows the necessity of studying musicality in autism, outside the context of domain-general cognitive models of talent.

Changing Concepts of Autism

Writing on the history of autism, Evans (2013) [17, 18] has described conceptual changes that have direct relevance to questions about musicality and other arts-based strengths in autism. She describes how, during the 1950s, the term autism was associated with 'hallucinatory dreamlike imaginary thought', and it was only during the 1960s, when attention was focused on distinguishing autism and childhood schizophrenia, that autism became associated with an absence of imagination and creativity. Although Kanner had originally described difficulties in 'affective contact', these characteristics were increasingly conceptualised as 'cognitive impairments' in social-communication skills. During the 1970s and 1980s Premack and Woodruff had studied social cognition in chimpanzees and young children [19, 20] and coined the term 'Theory of Mind' (ToM), to describe an ability to attribute autonomous mental states to oneself and others in order to understand and predict actions. From the mid-1980s the ToM deficit hypothesis [21] became the dominant theoretical framework within which 'social-communication impairments' in autism were studied and explained.

In 2012, Milton [22] published his account of the 'double empathy problem', and this motivated a new approach to conceptualising and studying autism. Findings from empirical studies showing that autistic people understand each other's thoughts and feelings [23], communicate more effectively with each other than with neurotypical people [24], and derive considerable benefits from engaging with other autistic people [25] challenged theories proposing that autistic people lack a theory of mind, or avoid social engagement. A second strand of research supporting Milton's account focused on exploring neurotypical perceptions of, and responses to, autistic people. For example, in a study by Sheppard, Pillai, Wong, Ropar and Mitchell (2016) [26], neurotypical adult participants were shown videotapes of autistic and neurotypical people and told that they were responding to jokes, compliments, stories or periods of waiting.

After viewing the videos the participants were asked to (1) identify the events that precipitated the responses in the individuals they observed, (2) evaluate their levels of expressivity and (3) provide verbal descriptions of their responses. The participants, who were unaware of the diagnostic status of the individuals shown in the videos, were less able to match responses with the events that had elicited them (e.g. jokes or compliments) for autistic compared with neurotypical individuals. This difficulty did not reflect differences in expressivity as the participants' ratings on this measure did not differ across groups. However, in their verbal descriptions of the responses they had observed, the participants made more frequent references to the internal mental states of autistic than neurotypical responders. This study provided clear evidence for neurotypical difficulties in understanding autistic people, and this was reflected in the increased use of mentalising language to describe the autistic people's actions.

Although the ToM account initially described an 'impairment' in understanding the thoughts, feelings and beliefs of others, it was later assumed that these 'difficulties' generalised to 'impairments' in self-knowledge [27]. Writing on this problem, Yergeau (2018) [28] has described how this claim 'disputes autistic people's autonomy, devalues their self-determination, and discredits their credibility'. In an appraisal of early autism research Bottema-Beutel, Kapp, Sasson, Gernsbacher, Natri and Botha (2023) [29] have argued that it has had a 'history of false leads in part because of unexamined ableist[1] ideologies that undergird research framings and interpretations of evidence'. They criticise the idea of an ideology-free science, questioning the 'scientific accuracy' of research that disregards the views and lived experiences of autistic people. A small but increasing body of research has explored autistic peoples' experiences of musical engagement, and some of this work will be described in the following section.

---

[1] In their article Bottema-Beutel and colleagues define ableism as 'a system of discrimination against people perceived to be disabled, based on socially constructed views of normalcy, productivity, desirability, intelligence, excellence and fitness'.

The Uses of Music in Everyday Life in Autism

Research exploring the utility of music in everyday life was described in Chap. 2. For example, a large-scale study by Schäfer and colleagues (2013) [30] addressed this question and showed that neurotypical children, adolescents and adults listen to music in order to regulate arousal and mood, to achieve self-awareness and as an expression of social relatedness. This has also been explored in studies of autistic adults and adolescents. In the earliest of these studies, Allen, Hill and Heaton (2009) [31] interviewed a group of 12 autistic adults and found that the majority used music to regulate arousal and mood, with half actively using music for therapeutic purposes. As will be further discussed, levels of anxiety and depression are significantly higher in autistic than in non-autistic populations and the therapeutic power of music was highlighted by several participants. For example, one described how 'when I have been feeling depressed, I have listened to certain music and I would claim the music healed me'. A second participant described how 'with autism you tend to feel dead a lot of the time—music is the key that unlocks the emotions'. The second most important theme to emerge in this study was aesthetic experience, with participants describing intellectual and emotional experiences that arose in direct response to the music. The importance of music for a sense of belonging or social relatedness was described by several participants. For example, one participant told how 'I'm reminded when I hear music, of people who like it… my sister likes particular music, when I listen to that I'm reminded of her'. Another person noted, 'what started me off collecting records was we had a friend, and he introduced me to Beano's in Croydon which is a second-hand record shop, and it started off from there'.

In a later study, Korosec, Osika and Boiner-Horwitz (2022) [32] asked 13 autistic adults to describe the roles that music played in their lives. Themes of wellbeing, identity and self-development, connectivity and negative experiences emerged in the study. Examples of wellbeing and coping involved the use of music to manipulate emotions and physiological arousal. Participants described how 'It can energise my whole body or it can put me practically to sleep' and 'Even if the music does not calm me, it makes the experience more pleasant. [...] It can change my agitated state [...] to something more neutral or even positive'. The connectivity theme comprised three subthemes: connecting, sharing experiences and a sense of belonging. The importance of concerts in providing opportunities for connecting with others and expanding one's social circle was described. Shared experiences included working with others towards a common goal during choir practice, spontaneously harmonising with other people and sharing ideas about music with them. A sense of belonging was eloquently described by one participant who told how 'with certain songs, I just feel connected to everybody, I don't know how to explain it'. The responses grouped in the second theme, identity and self-development, highlighted the importance of music for individual creativity and self-competence. Some of the participants painted, wrote, composed, played musical instruments, sang or danced, and according to one participant, 'It is probably precisely in this [creative] process that one gets to be a bit free, like ... to have one's own space'. Participants used music to stimulate creativity in other domains. One participant described using music for her writing: 'It can be motivating and inspiring [listening to music] and can prevent you from getting stuck in a text'. Another participant who paints described how 'I use music to get my creative process going'. The importance of music for creativity was also highlighted in a third qualitative study of musical experience in autistic adults. Matsuno, Auzenne and Chooskie (2021) [33] studied the use of music in a group of autistic creatives working in a video game development lab. In this study the participants described how they engaged with music-listening technologies, personalising their listening experiences to maximise their creativity and to meet their needs when performing everyday activities.

Korosec and colleagues identified factors that gave rise to negative responses to music in autistic adults. For example, some individuals reported that particular musical genres, or the sound quality and/or volume levels within pieces of music, could cause considerable discomfort. Disturbances in sensory processing are included in current diagnostic criteria for autism, and hyperacusis, characterised by a decreased tolerance for or extreme sensitivity to sound, has been widely reported in empirical studies [34]. It was clear from the participants' reports that when sounds are perceived as aversive and uncontrollable, they can create significant levels of stress. According to one participant, 'It is difficult to explain [...] It is like a-room-full-of-snakes uncomfortable'. Participants also reported being overwhelmed by the intensity of the emotions that some music evoked. For example, one participant who had attended a live performance of Tchaikovsky's *Pathétique* symphony said, 'There were a few times when it was almost too difficult to stay at a concert because I was so moved by the music'. Several other participants described the potentially overpowering affective qualities of music, with one reporting, 'Then I think, "Now I should really listen to something else so that I don't intensify the emotion unnecessarily"'. Other participants described how music could take over, causing them to neglect other important tasks, and the experience of 'crashing' after listening to powerfully engaging or emotional music.

Venter, Morelli and Erasmus (2023) [35] explored the role of music listening experiences in three adults and reported similar findings to those reported in the earlier studies described. The participants used music to cope with auditory sensitivities, for self-regulation, for finding connection and as a highly valued activity. Finally, in a study of 11 autistic adolescents and young adults, Kirby and Burland (2021) [36] noted that participants' musical interests and engagement spanned a wide range of musical genres. Participants reported on the importance of music for cognitive functions, for example to motivate and enhance activities and to inform routines. They also described music's importance for mood management, emotion regulation, decoding emotions and expressing feelings. Music was valued for its role in the search for personal identity and for enhancing opportunities for developing new and pre-existing relationships with others.

The results from these first-person accounts show that music serves multiple life-enhancing functions for autistic people. Amongst these functions, some may be unique, or at least utilised more frequently by autistic compared with neurotypical people. For example, some participants discussed the problem of sensory disturbances and the ways they used music to cope with them. However, despite this potential difference, comparison of the five studies described above and the findings reported in Schäfer et al. suggests striking similarities in the way that autistic and neurotypical people value music and utilise it in their everyday lives.

The importance of music for modulating mood must be considered in the context of what is known about differences in levels and experiences of anxiety and stress in neurotypical and autistic people. Krause and colleagues (2023) [37] investigated the types of stressors that elicited musical engagement in a large sample (n = 553) of adults living in Australia, Malaysia and the US. The results showed that participants used music to cope with social and work-related stressors more than other everyday stressors (financial, performance responsibilities, daily displeasures) identified in the study. They also reported that individuals who listened to music for emotion-/problem-oriented and avoidance/disengagement reasons were more likely to use music to cope with everyday stressors. According to research collated by the UK-based National Autistic Society, around half of all autistic people experience high levels of anxiety on a regular basis. In a study by McQuaid, Weiss, Said, Pelphrey, Lee and Wallace (2022) [38] this difficulty was most strongly associated with reduced independence in activities of daily living and poorer subjective quality of life. In a community-based participatory research study Raymaker and colleagues (2020) [39] investigated autistic meltdown, a widely reported condition about which little was previously known. The authors reported that autistic meltdown, manifested in chronic exhaustion, loss of skills and reduced tolerance to environmental stimulus, results from an accumulation of life stressors and barriers to support systems.

In addition to utilising music for day-to-day coping, autistic adults and adolescents used music for personal development, creativity and motivation, and these factors will be explored in more detail in the following chapters. A finding that emerged most strongly in the study by Allen and colleagues was that autistic adults valued music for its aesthetic qualities and the emotional experiences these gave rise to. Juslin's (2013) model of musical emotions includes a multi-stage account of aesthetic responses to music. In an initial stage of listening, auditory information is pre-classified as 'art', and its perceptual qualities (e.g. sensory and Gestalt components), higher-order qualities (style, performance, composer) and emotion inputs (visual imagery, autobiographical memory, associative learning, etc.) are integrated in an aesthetic filtering system. This filtering system is shaped by an individual's personal concepts of beauty, skills, novelty, style, message, expression and emotion. Kirby and Burland (2021) reported that the participants in their study engaged with music from a wide range of genres and this reflects the intensely personal nature of musical preferences.

The importance of music for a sense of belonging and relatedness was reported by many autistic participants in the studies. In the words of a participant in the study by Korosec and colleagues, 'with certain songs, I just feel connected to everybody, I don't know how to explain it'. In the study of neurotypical people carried out by Schäfer and colleagues (2013) music was important for social relatedness, but this function was less important than music's utility in regulating arousal and mood. Evolutionary accounts of music emphasise the importance of interpersonal and group functions, and these findings suggest a dissociation between music's current functions and functions believed to have driven its evolution [40, 41]. However, a recent study carried out by Schäfer, Saarikallio and Eerola (2020) [42] established links between an individual's embodied experience of social-emotional relationships and their uses of music in everyday life. In the study, the authors used mood induction and guided imagery to test the hypothesis that specific types of music may function as an empathic friend. The results confirmed the hypothesis, showing that participants experienced improvements in mood and empathy, and reduced feelings of loneliness after listening to self-selected comforting music. In the study by Allen and colleagues autistic adults described how they associated music with valued friends and family members and used music to induce positive emotion.

Evidence from first-person accounts shows that autistic people value music for multiple and complex reasons, and for some individuals music takes on a central role in everyday life. In the following chapters I will explore musicality based on the first-person accounts of autistic people for whom this is the case.

References

1. Trehub, S. (2001). Musical predispositions in infancy. *Annals of the New York Academy of Sciences, 930*, 1–16. https://doi.org/10.1111/j.1749-6632.2001.tb05721.x
2. Blood, A. J., & Zatorre, R. J. (2001). Intensely pleasurable responses to music correlate with activity in brain regions implicated in reward and emotion. *Proceedings of the National Academy of Sciences of the United States of America, 98*(20), 11818–11823. https://doi.org/10.1073/pnas.191355898. PMID: 11573015; PMCID: PMC58814.
3. Happé, F., & Vital, P. (2009). What aspects of autism predispose to talent? *Philosophical Transactions of the Royal Society of London. Series B, Biological Sciences, 364*(1522), 1369–1375. https://doi.org/10.1098/rstb.2008.0332. PMID: 19528019; PMCID: PMC2677590.
4. Happé, F., & Frith, U. (2006). The weak coherence account: Detail-focused cognitive style in autism spectrum disorders. *Journal of Autism and Developmental Disorders, 36*(1), 5–25. https://doi.org/10.1007/s10803-005-0039-0. PMID: 16450045.
5. Baron-Cohen, S., & Lombardo, M. V. (2017). Autism and talent: The cognitive and neural basis of systemizing. *Dialogues in Clinical Neuroscience, 19*(4), 345–353. https://doi.org/10.31887/DCNS.2017.19.4/sbaroncohen. PMID: 29398930; PMCID: PMC5789212.
6. Mottron, L., Bouvet, L., Bonnel, A., Samson, F., Burack, J. A., Dawson, M., & Heaton, P. (2013). Veridical mapping in the development of exceptional autistic abilities. *Neuroscience and Biobehavioral Reviews, 37*(2), 209–228. https://doi.org/10.1016/j.neubiorev.2012.11.016. Epub 2012 Dec 5. PMID: 23219745.
7. Heaton, P., Hermelin, B., & Pring, L. (1999). Can children with autistic spectrum disorders perceive affect in music? An experimental investigation. *Psychological Medicine, 29*(6), 1405–1410. https://doi.org/10.1017/s0033291799001221. PMID: 10616946.
8. Heaton, P., Allen, R., Williams, K., Cummins, O., & Happé, F. (2008). Do social and cognitive deficits curtail musical understanding? Evidence from autism and Down syndrome. *British Journal of Developmental Psychology, 26*(2), 171–182. https://doi.org/10.1348/026151007X206776
9. Quintin, E. M., Bhatara, A., Poissant, H., Fombonne, E., & Levitin, D. J. (2011). Emotion perception in music in high-functioning adolescents with autism spectrum disorders. *Journal of Autism and Developmental*

Disorders, 41(9), 1240–1255. https://doi.org/10.1007/s10803-010-1146-0. PMID: 21181251.
10. Sivathasan, S., Dahary, H., Burack, J. A., & Quintin, E.-M. (2023). Basic emotion recognition of children on the autism spectrum is enhanced in music and typical for faces and voices. *PLoS One, 18*(1), e0279002. My 1999paper.
11. Kastner, M. P., & Crowder, R. G. (1990). Perception of the major/minor distinction: IV. Emotional connotations in young children. *Music Perception, 8*(2), 189–201. https://doi.org/10.2307/40285496
12. Macari, S., Milgramm, A., Reed, J., Shic, F., Powell, K. K., Macris, D., & Chawarska, K. (2021). Context-specific dyadic attention vulnerabilities during the first year in infants later developing autism spectrum disorder. *Journal of the American Academy of Child and Adolescent Psychiatry, 60*(1), 166–175. https://doi.org/10.1016/j.jaac.2019.12.012. Epub 2020 Feb 13. PMID: 32061926; PMCID: PMC9524139.
13. Huron, D. (2006). *Sweet anticipation: Music and the psychology of expectation.* The MIT Press.
14. Salimpoor, V. N., Zald, D. H., Zatorre, R. J., Dagher, A., & McIntosh, A. R. (2015). Predictions and the brain: How musical sounds become rewarding. *Trends in Cognitive Sciences, 19*(2), 86–91. https://doi.org/10.1016/j.tics.2014.12.001
15. Juslin, P. N., & Västfjäll, D. (2008). Emotional responses to music: The need to consider underlying mechanisms. *Behavioral and Brain Sciences, 31*(5), 559–575.
16. Juslin, P. N. (2013). From everyday emotions to aesthetic emotions: Towards a unified theory of musical emotions. *Physics of Life Reviews, 10*(3), 235–266. https://doi.org/10.1016/j.plrev.2013.05.008. Epub 2013 May 29. PMID: 23769678.
17. Evans, B. (2013). How autism became autism: The radical transformation of a central concept of child development in Britain. *History of the Human Sciences, 26*(3), 3. https://doi.org/10.1177/0952695113484320
18. Evans, B. (2017). *The metamorphosis of autism: A history of child development in Britain.* Manchester University Press. PMID: 28654228.
19. Premack, D., & Woodruff, G. (1978). Does the chimpanzee have a theory of mind? *Behavioral and Brain Sciences, 1*(4), 515–526. https://doi.org/10.1017/S0140525X00076512

20. Premack, D., & Premack, A. J. (1995). Levels of causal understanding in chimpanzees and children. In J. Mehler (Ed.), *Cognition on cognition* (p. xi, 486). The MIT Press.
21. Baron-Cohen, S., Leslie, A. M., & Frith, U. (1985). Does the autistic child have a "theory of mind"? *Cognition, 21*(1), 37–46. https://doi.org/10.1016/0010-0277(85)90022-8. PMID: 2934210.
22. Milton, D. (2017). *A mismatch of salience* (pp. 883–887). Pavilion.
23. Heasman, B., & Gillespie, A. (2019). Neurodivergent intersubjectivity: Distinctive features of how autistic people create shared understanding. *Autism, 23*(4), 910–921. https://doi.org/10.1177/1362361318785172. Epub 2018 Aug 3. PMID: 30073872; PMCID: PMC6512057.
24. Crompton, C. J., Ropar, D., Evans-Williams, C. V., Flynn, E. G., & Fletcher-Watson, S. (2020). Autistic peer-to-peer information transfer is highly effective. *Autism, 24*(7), 1704–1712. https://doi.org/10.1177/1362361320919286
25. Crompton, C. J., Hallett, S., Ropar, D., Flynn, E., & Fletcher-Watson, S. (2020). 'I never realised everybody felt as happy as I do when I am around autistic people': A thematic analysis of autistic adults' relationships with autistic and neurotypical friends and family. *Autism, 24*(6), 1438–1448. https://doi.org/10.1177/1362361320908976
26. Sheppard, E., Pillai, D., Wong, G. T. L., Ropar, D., & Mitchell, P. (2016). How easy is it to read the minds of people with autism spectrum disorder? *Journal of Autism and Developmental Disorders, 46*, 1247–1254.
27. Gernsbacher, M. A., & Yergeau, M. (2019). Empirical failures of the claim that autistic people lack a theory of mind. *Archives of Scientific Psychology, 7*(1), 102–118. https://doi.org/10.1037/arc0000067
28. Yergeau, M. (2018). *Authoring autism: On rhetoric and neurological queerness*. Duke University Press.
29. Bottema-Beutel, K., Kapp, S. K., Sasson, N., Gernsbacher, M. A., Natri, H., & Botha, M. (2023). Anti-ableism and scientific accuracy in autism research: A false dichotomy. *Frontiers in Psychiatry, 14*, 1244451. https://doi.org/10.3389/fpsyt.2023.1244451. PMID: 37743979; PMCID: PMC10514488.
30. Schäfer, T., Sedlmeier, P., Städtler, C., & Huron, D. (2013). The psychological functions of music listening. *Frontiers in Psychology, 4*, 511. 10.3.
31. Allen, R., Hill, E., & Heaton, P. (2009). The subjective experience of music in autism spectrum disorder. *Annals of the New York Academy of Sciences, 1169*, 326–331. https://doi.org/10.1111/j.1749-6632.2009.04772.x. PMID: 19673801.389/fpsyg.2013.00511.

32. Korošec, K., Osika, W., & Horwitz, E. (2022). "It is more important than food sometimes"; meanings and functions of music in the lives of autistic adults through a hermeneutic-phenomenological Lense. *Journal of Autism and Developmental Disorders, 54*, 1–13. https://doi.org/10.1007/s10803-022-05799-2
33. Matsuno, M., Auzenne, D., & Chukoskie, L. (2020). "All bets are off": Flexible engagement with music-listening technologies by autistic adults. *Psychology of Music, 49*, 030573562097103. https://doi.org/10.1177/0305735620971037
34. Danesh, A. A., Howery, S., Aazh, H., Kaf, W., & Eshraghi, A. A. (2021). Hyperacusis in autism spectrum disorders. *Audiol Res., 11*(4), 547–556. https://doi.org/10.3390/audiolres11040049. PMID: 34698068; PMCID: PMC8544234.
35. Venter, F., Morelli, J., & Erasmus, E. (2023). Understanding the lived music listening experiences of adults on the autism spectrum. *Psychology of Music, 51*(3), 971–985. https://doi.org/10.1177/03057356221126201
36. Kirby, M. L., & Burland, K. (2022). Exploring the functions of music in the lives of young people on the autism spectrum. *Psychology of Music, 50*(2), 562–578. https://doi.org/10.1177/03057356211008968
37. Krause, A. E., Scott, W. G., Flynn, S., Foong, B., Goh, K., Wake, S., Miller, D., & Garvey, D. (2023). Listening to music to cope with everyday stressors. *Musicae Scientiae, 27*(1), 176–192. https://doi.org/10.1177/10298649211030318
38. McQuaid, G., Weiss, C., Said, A., Pelphrey, K., Lee, N., & Wallace, G. (2022). Increased perceived stress is negatively associated with activities of daily living and subjective quality of life in younger, middle, and older autistic adults. *Autism Research, 15*, 1535. https://doi.org/10.1002/aur.2779
39. Raymaker, D. M., Teo, A. R., Steckler, N. A., Lentz, B., Scharer, M., Delos Santos, A., Kapp, S. K., Hunter, M., Joyce, A., & Nicolaidis, C. (2020). "Having all of your internal resources exhausted beyond measure and being left with no clean-up crew": Defining autistic burnout. *Autism Adulthood, 2*(2), 132–143. https://doi.org/10.1089/aut.2019.0079. Epub 2020 Jun 10. PMID: 32851204; PMCID: PMC7313636.
40. Mehr, S. A., Krasnow, M. M., Bryant, G. A., & Hagen, E. H. (2021). Origins of music in credible signaling. *Behavioral and Brain Sciences, 44*, e60. https://doi.org/10.1017/S0140525X20000345
41. Savage, P. E., Loui, P., Tarr, B., Schachner, A., Glowacki, L., Mithen, S., & Fitch, W. T. (2021). Music as a coevolved system for social bonding.

Behavioral and Brain Sciences, 44, 1–22. https://doi.org/10.1017/S0140525X20000333. Available at https://centaur.reading.ac.uk/95527/, ISSN 1469-1825

42. Schäfer, K., Saarikallio, S., & Eerola, T. (2020). Music may reduce loneliness and act as social surrogate for a friend: Evidence from an experimental listening study. *Music & Science, 3*. https://doi.org/10.1177/2059204320935709

5

First-Person Accounts of Musical Talent in Autistic Adults

Abstract In this chapter I present brief autobiographical accounts of a group of six autistic musicians. These individuals have generously shared their insights into their musical experiences and journeys, and their beliefs about how autism is conceptualised. My correspondence with some of these musicians stretches back over several years, with initial contact resulting from our shared interests in music and musical experience. Interview material included in this and the following two chapters was collected in multiple person-to-person meetings, online meetings and via e-mail.

Keywords Musical talent · Musical engagement · Autism Quotient Scale · Diagnostic criteria · Factors

In Chap. 4 studies exploring musical engagement from the first-person perspectives of autistic adults and adolescents were described. The findings from these studies showed that musical engagement is powerfully influenced by an appreciation of music's emotional and aesthetic qualities and its benefits for mood modulation, creativity, cognition and social affiliation. This work provides an important new context for understanding the emergence and development of musical talents in autistic individuals for whom musical experiences are particularly intense and rewarding.

Musical Journeys

Sarah

Sarah was diagnosed with autism in adulthood after she had relocated from Asia to study in the UK. As a very young child Sarah had shown exceptionally strong musical and intellectual abilities and had been admitted to a government-sponsored programme for child prodigies. At five years Sarah began to study violin and piano and made extremely rapid progress, completing all of her formal assessments by the time she was 10 years old. Looking back on her early childhood she reports that she can remember the first music she learned much better than she can remember the first things she talked about. However, despite the importance of these early experiences, she reports that she didn't really enjoy playing classical music. In sharp contrast to her later experiences as an improvising jazz and funk musician, she did not find classical music intellectually or emotionally engaging and never experienced flow when playing classical music. As a child, Sarah experienced some exposure to jazz composition and improvisation when she attended a group music programme, but reports that her most powerful and formative early musical experiences occurred when she was playing the piano in her church band and needed to focus on the musical interactions with the other band members. Sarah believes that her early classical training was overly focused on the development of individual technical skills, and she contrasts this with the experiences of children who develop their musical skills outside of the Western classical music tradition. Citing the example of the African American gospel tradition, she described how learning to play with others is the starting point for musical development. Growing up in an environment that was largely dominated by Western classical music, Sarah's musical journey took a very different direction when she was old enough to explore different musical genres herself. Whilst she still occasionally plays in a classical orchestra, her overriding passion is for jazz and funk, and despite the demands of her work as a scientist, she is an active musician, playing frequently in professional ensembles.

Elisabeth

Elisabeth is a professional orchestral and solo cellist. Her own parents were trained classical musicians and her early musical environment was very rich. Elisabeth recalls an early formative experience that occurred when her father was rehearsing for a concert with the principal cellist of the Gothenburg Symphony Orchestra. The rehearsals took place in Elisabeth's home and when they ended the cellist spent some time playing children's songs for her. This sparked an interest in the instrument that became apparent to her parents when they found her sitting on a step playing a 'cello' that she had constructed from two long sticks. With the encouragement and assistance of the cellist, lessons were organised and by three and a half years she was studying cello on a 10th sized instrument built for very young children. In addition to studying the cello Elisabeth began to teach herself the piano, with help from her father when needed. Unlike Sarah, Elisabeth always found classical music emotionally and intellectually fulfilling. She recalls long summers happily spent mastering Bach's preludes and fugues and she reports prizing her two instruments for the different modes of expression they afforded ('like mummy and daddy'). When Elisabeth auditioned for the Gothenburg Music Academy for entry onto the undergraduate programme her skills on both instruments were recognised. However, despite her love of the piano and the piano repertoire, Elisabeth's long-held ambition was to play in a symphony orchestra and she chose to specialise in the cello. During her undergraduate years she was taught by excellent cellists who enabled her to acquire the work ethic that is so essential in this demanding profession, and as a postgraduate student, Elisabeth was assigned to a teacher who became a great source of inspiration. She reports how this teacher had a particularly direct and detailed way of communicating and was able to understand and respond to Elisabeth's personal style of learning. After she completed her Master's degree, Elisabeth won a prestigious musician of the year award and joined the Amsterdam Concertgebouw Orchestra as an apprentice musician. Following this she held positions in the Netherlands Philharmonic Orchestra and the London Philharmonic Orchestra where she served for 10 years. Currently Elisabeth is based in Sweden, where she works as a cello soloist, chamber musician and teacher. She also holds a position in the Swedish church and is developing her skills as a pianist, organist, choral conductor and administrator. Elisabeth is an active autism advocate and her work in this area will be briefly described in Chap. 6.

Will

Will is a graduate of a London music academy and is in the early stages of developing his career as a session guitarist and teacher. Will's family report that he showed an intense interest in music from a very early age. By the age of three he had developed a passion for 'Dead Ringer for Love' written by Jim Steinman and recorded by Meatloaf. His interest in the guitar was first stimulated when he learned about a mini-guitar a friend had received for Christmas. Once Will got his own guitar and began having lessons at the age of six, his passion for this instrument and the music he could play on it became a powerfully motivating aspect of his everyday life. For example, he records how his motivation to play 'Sweet Child O' Mine' by Guns N' Roses resulted in him mastering a very difficult chord shape that his teacher had assured him was impossible for a child with such small hands. Will's musical preferences are constantly evolving. His early, most powerful interest was in classic rock and he has described how his musical development has been most strongly influenced by Black Sabbath, Zeppelin, ACDC, Foo Fighters and Simple Minds. More recently he has become interested in jazz and rhythm and blues. Will was diagnosed with autism as a young child and has always experienced severe sensory disturbances. On what is already an exceptionally demanding career path, these difficulties impose multiple additional challenges.

Although Will has been active in developing strategies that enable him to cope with the sensory disturbances he sometimes experiences in rehearsal and performance settings, he has found that musician colleagues are not always able to understand the nature and impact of these problems. Will believes that the experience of performing with other musicians has been crucial for his own musical development, although communication breakdowns, directly resulting from sensory disturbance, present ongoing challenges. Currently Will is most interested in playing rock and alternative rock and extracts of his playing can be accessed via the following link: https://youtu.be/j-_BojFvgPY?si=vLFyiomMQe1m8Zw5.

Andrew

Andrew is a composer, a musicologist, a creative technologist and a professor in a UK university. Born in 1957, he grew up in the period when children with strong intellectual and language skills were not diagnosed with autism. Moreover, sensory disturbance, Andrew's most troubling early difficulty, was not included in diagnostic criteria until the publication of DSM-5 in 2013 [1]. As a result of these and other factors, he was not diagnosed with autism until he was 60 years old. Andrew began to study music as a very young child and it quickly became his primary preoccupation. From an early age he actively explored different musical genres and at 11 years, began to compose his own music. One motivation for this was that he became aware that he could express things in music that he could not express in words. Andrew took piano lessons from an early age, and the school he attended had an extensive record library which he was able to access. From this early period Andrew's exploration of different musical genres has been an important aspect of his musical journey. Andrew's perception of his early auditory world was strongly shaped by hyperacusis, a condition characterised by a reduced tolerance and increased sensitivity to everyday sounds in the environment [2]. He reports how he gradually became aware that other people's perceptions of their auditory world did not include sounds emitted from refrigerators and overhead lights. His perception of pitch was exceptionally fine-grained and this meant that he perceived dissonance in musical sounds (e.g. tonic triads) that are categorised as consonant within an 'equal temperament' system of pitch tuning. As will be described in Chap. 7, Andrew's interest in the work of composers who creatively explored tonality enabled him to understand his own perception of dissonance and consonance and adapt to a tuning system he found fundamentally unnatural. Andrew has synaesthesia, and like many composers with this condition, this has shaped and enriched his experiences of sound and influenced his creative work. An example of Andrew's creative work exploring the relationship between sound and colour will be described in Chap. 7.

Stéphane

Stéphane began to take weekly piano lessons when she was seven years old, and whilst she has continued to study music formally throughout her childhood, adolescence and into adulthood she believes that this had far less influence on her musical development than her own exploration of musical instruments and sounds. She recalls her early passion for musical exploration, reporting that the keyboard served as a 'human/sound interface', the guitar was an excellent medium for improvisation and exploration, and the creative potential of percussion instruments was seemingly limitless. The importance of music as a creative aspect of Stéphane's early lived experience is vividly captured in many of her childhood memories. For example, she recalls an occasion when she assembled a selection of saucepans and other kitchen utensils, with the intention of giving a percussion recital to her 'audience' of stuffed animal toys. Children studying Western classical music typically learn musical notation and Stéphane found this extremely onerous. Although she did learn musical notation, her dissatisfaction with its utility as a representational system increased in line with her developing skills as a musician. She admits that she experienced satisfaction when she mastered a difficult musical score and could play it on the piano. But this was experienced as 'satisfaction from a job well done' rather than the 'true musical pleasure' of engaging in 'natural and effortless' musical exploration. Although the process of learning musical notation is challenging for many children, Stéphane's difficulties were exacerbated by her sensory sensitivities and atypical perceptual and cognitive processing. Stéphane has synaesthesia, and the influence this has had on her musical development will be discussed in Chap. 7. As an adult working in creative technology, Stéphane takes a multidisciplinary approach to her musical studies. She has completed undergraduate and post-graduate degrees in music and music-related subjects in the UK and Canada and is currently developing a PhD project.

James

James is a composer and a clarinetist who has recently completed a PhD in composition at a British University. None of James's family members played musical instruments but as a young woman their mother had trained and worked as a classical ballerina and they recall an early experience of dancing to a CD of Prokofiev's *Peter and the Wolf*. In their description of their development James contextualises their musicality in their uneven pattern of strengths and difficulties. As a child James was hyperlexic and very good at arithmetic and chess but found sports and group activities highly aversive. They experienced mild dyspraxia that affected their balance and co-ordination and they developed a 'crippling phobia' around group activities. At age seven James's mother insisted that they learn to play the recorder, from which they made a later transition to the clarinet. The most immediate effect of this shift was that they began to play music that was intellectually and emotionally satisfying. They report that the first piece they learned to play on the recorder was an eight-bar 'composition' consisting of repetitions of the note B in 4/4 time. After a short period of learning the clarinet they were able to play the 'cat' variation from Prokofiev's *Peter and the Wolf* and from this point they began to master the clarinet classical repertoire. James's clarinet teacher was a virtuosic saxophonist and in addition to teaching the clarinet and music theory, she encouraged them to explore the possibilities of their instrument and the different sounds they could make on it. They joined a jazz band organised by a professional jazz player and quickly rose to the position of first clarinet. Around this time, they took a family holiday in a house that had a piano and beginner piano music books and by the end of the holiday they had learned some scales and begun to compose piano pieces.

At the age of 13 James excelled at music and academic subjects and won a scholarship to a prestigious private school. However, they report that the school placed a low value on academic and artistic achievements and did not implement anti-bullying policies. The result of this was that James's early teenage years were marked by periods of severe stress-related illnesses alongside an increasing love of and capacity for musical performance and composition. They report how they worked to master challenging compositions within the clarinet repertoire and won a national competition when they were 13 years old. Support for their musical development during their adolescent period came from their piano accompanist and their piano teacher. They also took lessons with a composition teacher, and the influence of these sessions will be described in Chap. 7. James was diagnosed with Asperger's syndrome at a time when they were experiencing considerable distress. They write, 'There is a certain cruel irony that I only received a diagnosis out of catastrophe, and that all of my obvious social problems were ignored because I had success in other areas'. They describe how they continue to struggle with things neurotypical people find easy (social interactions) whilst showing excellence in the arts and academic subjects.

Although the autism+music experts' accounts revealed differences in musical interests, modes of engagement and trajectories of musical development, all described early and intense interests that have been documented in studies of child prodigies [3, 4]. In a study exploring the factors associated with prodigious musical achievements, Marion-St-Onge, Wiess, Sharda and Peretz (2020) [5] compared groups of musicians who had and had not met criteria for prodigy status in childhood. Motivated by the Multifactorial Gene-Environment Interaction Model [6], the authors explored group differences on tests of intelligence, personality, motivation, reward and levels of practice. The results showed that groups did not differ on any of the measures of intelligence or personality, and the findings suggested that qualitative differences in experiences of flow and differences in patterns of practise had distinguished the groups at early stages of development.

The study by Marion-St-Onge and colleagues has important implications for understanding exceptional musicality in autism. Two of the musicians in the prodigy group had been diagnosed with autism in childhood, and the authors measured levels of autism traits in the whole sample using the Autism Quotient Scale (AQ) [2]. Although levels of AQ traits were high in the two autistic prodigies and one other musician, comparison of musician and non-musician groups failed to reveal differences on total AQ scores, or on scores from the five factors that link with diagnostic criteria for autism. On the basis of these findings the authors ruled out a link between exceptional musicality and autism. However, in the following chapter this link will be re-examined in the context of alternative accounts of autism and the nature of musical experience.

References

1. American Psychiatric Association. (2013). *Diagnostic and statistical manual of mental disorders* (5th ed.). American Psychiatric Association. https://doi.org/10.1176/appi.books.9780890425596
2. Baron-Cohen, S., Wheelwright, S., Skinner, R., Martin, J., & Clubley, E. (2001). The autism spectrum quotient: Evidence from asperger syndrome/high functioning autism, males and females, scientists and mathematicians. *Journal of Autism and Developmental Disorders, 31*, 5–17. https://doi.org/10.1023/A:1005653411471
3. Winner, E. (1996). *Gifted children: Myths and realities.* Basic Books.
4. Winner, E. (2000). The origins and ends of giftedness. *American Psychologist, 55*(1), 159–169. https://doi.org/10.1037/0003-066X.55.1.159
5. Marion-St-Onge, C., Weiss, M., Sharda, M., & Peretz, I. (2020). What makes musical prodigies? *Frontiers in Psychology, 11*, 566373. https://doi.org/10.3389/fpsyg.2020.566373
6. Ullén, F., Hambrick, D. Z., & Mosing, M. A. (2016). Rethinking expertise: A multifactorial gene-environment interaction model of expert performance. *Psychological Bulletin, 142*(4), 427–446. https://doi.org/10.1037/bul0000033. Epub 2015 Dec 21. PMID: 26689084.

6

Musical Journeys 1: Monotropism, Flow and Musicality

Abstract In this chapter Sarah's and Elisabeth's musical experiences, and similarities and differences in their musical preferences and modes of engagement, will be explored. These accounts will be contextualised in theoretical models derived from autistic people's lived experiences as well as in research documenting rewards associated with musical engagement.

Keywords Monotropism • Flow • Musicality • Musical preferences

In addition to her work as a professional musician, Elisabeth is an autism advocate working within the Neurodiversity Paradigm. Within her working environments, the majority of Elisabeth's colleagues are neurotypical and an important aim in her advocacy work is to promote neurotypical understanding of neurocognitive differences that influence the way autistic people think, learn, communicate and experience the world. Elisabeth rejects traditional diagnostic conceptualisations of autism and endorses the monotropism theory of autism, which she believes maps most closely onto her own lived experience as an autistic person.

Monotropism

The theory of monotropism, first proposed by Murray, Lesser and Lawson in 2005 [1], describes how an individual's attention may be broadly allocated across multiple interests (polytropism) or focused more intensely on a small range of interests (monotropism). Although this model does not assume that the latter characteristic is unique to autism, it does propose that autistic people are at the extreme end of this continuum of attentional styles. A major strength of this theory is that it accounts for both strengths and difficulties in autism within a single non-pathologising account. Writing on monotropism and autism, Murray (2018) [2] has described how the demands within the social world are most easily met by individuals with broadly distributed attention. Writing on differences between autistic and neurotypical people he writes, 'Many social differences are sensory differences at root. Being unable to process multiple channels of input most of the time makes the combination of spoken words, body language and eye contact tremendously challenging'. However, he also describes how information that captures an individual's interests is processed within an 'attentional tunnel', where it is integrated and elaborated and gives rise to considerable intellectual and emotional rewards.

Flow

Deep immersion in activities of personal interest is powerfully linked with experiences of flow. The concept of flow was first proposed by Csikszentmihalyi (1975) [3] to describe a state in which a person's total immersion in an activity engenders feelings of absorption, energy and enjoyment. According to Csikszentmihalyi's early account, flow is multifaceted and encompasses nine key dimensions. In order to achieve flow an individual (1) must experience an equal balance between challenges and skills levels, (2) be completely absorbed in the task, and (3) have a clear sense of purpose and an awareness of what will happen next. During periods of flow the individual experiences (4) direct and immediate feedback so that s/he is able to constantly adjust her/his reactions, (5) a very high level of focus on the task in hand, (6) a strong sense of personal control, (7) a loss of self-consciousness and (8) a distorted sense of time. The ninth dimension, autotelic experience, refers to feelings arising in response to activities that are performed entirely for their own sake. Although this dimension is an important component in Csikszentmihalyi's initial conceptualisation of flow states, Csikszentmihalyi and Csikszentmihalyi (1988) [4] later suggested that some individuals possess psychological traits that increase their propensity to experience flow. They described an 'autotelic' personality type, suggesting that they may possess an ability to achieve a rewarding balance between the 'play' of challenge finding and the 'work' of skill building (1993, p.80). This component of flow can therefore be conceptualised as a state, which is experienced for a period of time, or a trait which is a fundamental aspect of an individual's personality.

Autobiographical accounts written by autistic adults show that experiences of flow are a pervasive aspect of everyday life. Writing on flow, Milton (2017) [5] has described how this may be mischaracterised as a 'tuning out' of the social world, rather than a deep immersion in activities that capture interests and trigger pleasurable experiences of flow. He describes how aspects of flow states, for example clear goals, immediate feedback, the fusing of action and awareness, reduced anxiety around failure and the sense of control and achievement, bring intrinsic satisfaction. In a recent study carried out by Rapaport, Clapham, Adams, Lawson, Porayska-Pomsta and Pellicano (2023) [6] autistic adults described the experience of being deeply immersed in tasks that capture their interests, and reported how they sometimes find it difficult to achieve a balance between these life-enhancing experiences and the demands of their everyday lives. The experience of being interrupted during flow states can be very distressing for autistic people. Elisabeth tells how 'I know now to set an alarm when engaging in my interests if I have to be somewhere, but it is important that I set more than one. I need a warning to start making the cognitive journey out of my attention tunnel. Sometimes it's very hard to stop what you are doing'.

Flow and Music

Csikszentmihalyi (1990) [7] has described how music 'helps organise the mind that attends to it, and therefore reduces psychic entropy, or the disorder we experience when random information interferes with goals. Listening to music wards off boredom and anxiety, and when seriously attended to it can induce flow experiences' (p109). Csikszentmihaly's account of music's functions powerfully resonates with autistic people's descriptions of the ways they use music in their everyday lives.

Although research carried out with neurotypical participants shows that flow experiences are associated with multiple activities, it also suggests that the relationship between music and flow is particularly powerful. In a review of 56 studies investigating musically induced flow Tan and Sin (2021) [8] reported that most of this work used self-report questionnaires that probe the different dimensions of flow described by Csikszentmihalyi. The results from these studies have shown that flow is experienced in response to multiple musical activities by individuals both with and without musical training. For example, studies have reported that flow can be experienced whilst listening to music, as well as when performing, composing, conducting and teaching music. It is experienced both within groups of musicians and between musicians and audiences. Flow is powerfully linked with creativity and an increased propensity for musical engagement, practice and achievement.

Monotropism and Communication in Different Contexts

The monotropism theory situates autistic strengths in the context of interests and rewards, and the autism+music experts have all described how musical activity has always been deeply engrossing and rewarding. The monotropism theory also provides an explanation for the difficulties autistic people may experience in some social contexts. Elisabeth has described how neurotypical, or polytropic, people are able to enjoy social 'chit-chat', for example in restaurants, because they are not 'thrown' by smells, sounds and visual stimuli in the immediate environment, or by rapid shifts from one topic to another in conversations. While several of the autism+music experts described how they experience difficulties engaging in 'social chit-chat', with some actively disliking it, they also described the satisfaction of talking with neurotypical people about topics of mutual interest. This suggests that shared interests, as well as sensory and other differences distinguishing autistic and neurotypical people, influence communication between them. Monotropism and flow theories [9] describe an autistic propensity to engage in deeply focused and rewarding activities, and research into musically induced flow shows that music elicits these behaviours and responses in neurotypical as well as in autistic people. The conceptualisation of autism as a condition characterised by 'impairments' in social and communication skills [10] has been powerfully challenged by Milton's double empathy account [11] and empirical evidence documenting neurotypical biases and difficulties in understanding autistic people [12, 13]. However, it is important to consider how empathic connections between autistic and neurotypical individuals may be influenced by the nature of their shared interests, particularly when this is music.

Diagnostic criteria for ASD in DSM-5 [10] include 'a failure of normal back-and-forth conversation, reduced sharing of interests, emotions, or affect, and a failure to initiate or respond to social actions'. However, Sarah writes, 'as an improvising musician I am constantly working out what to play with respect to the piece itself, what the rest of the band is doing, what the audience wants to hear etc. And this requires a lot of on-the-spot flexibility and a deep understanding of the underlying musical structures'. These interactions are all about communication, but in this case they are occurring in a context where the interests, motivations and emotions of all participating individuals are in close alignment.

In his account of communication in music, Cross (2014) [14] has described how 'interactive engagement with others through music provides the space for the sense to emerge that the attitudes and motivations of each interactant are being honestly communicated to each other and are in alignment'. He goes on to say 'we experience musical events and behaviours as having meanings that are both shared and yet are intensely personal and idiosyncratic, but any tension between mutual and personal meaning is neither expressed nor shared'.

Sarah has described how a shared appreciation of specific musical genres shapes social norms within musical cultures. Writing on the world of jazz, she reports that 'no one cares if you don't sit and socialise with others. As long as you can play so that you can make the others sound good then you'll still get booked. Jazz musicians are very tolerant in that respect—one doesn't need to have good social skills'. According to Sarah, acceptance within this world is primarily based on the individual's musical skill set, with reliability, punctuality and the ability to deal with the gear being added bonuses. Elisabeth has described similarly rewarding experiences working as a classical orchestral musician and these will be discussed later in the chapter.

Individual Differences in Musical Preferences

In Swanick and Tillman's (1986) [15] account of musical development the earliest *mastery* stage (0–4 years) is followed by successive stages in which an individual's musical outputs show increases in personal expressivity and creativity. During the adolescent period individuals begin to identify with musical communities and theorising about music becomes increasingly sophisticated and integrated in their broader understanding of themselves and their social world. Musical development is influenced by both communal and personal factors, and individual differences in personality will influence an individual's journey through the different stages of musical development.

As a child enrolled on a programme for musically gifted children Sarah had made rapid early progress on the violin and piano. However, she also excelled in mathematics and other academic subjects, and believes that her early approach to learning about music and these other subjects was very similar. As she was to later understand, her early approach to learning about music reflected her lack of engagement with the compositions she learned during this period. Describing her experiences of playing music by Beethoven, she recalls an absence of any emotional connection or understanding about what the composer was aiming to express in his music. Sarah's first experience of improvising with her church band motivated a dramatic shift in her understanding of what it is to be a musician,

6   Musical Journeys 1: Monotropism, Flow and Musicality

and she believes that this marked the true starting point of her musical development.

An equally important factor in Sarah's musical development was her search for music she found personally meaningful. Sarah is fascinated by the temporal qualities of music. She loves complex rhythms and the way these evolve over the course of a live performance. She described how 'my jazz playing colleagues (like me) are obsessed about this concept of groove, in which the music locks together and produces intense pleasure. It is quite intangible though and there are lots of micro-elements which contribute to it, for example, in funk—the use of ghost notes and their relative volume, in jazz—the amount of dottedness in a swing beat and how many milliseconds the bass player plays behind the drummer, or how straight/swung the pianist plays to create a sense of tension'. Sarah has also described the pleasure of performing with others. She tells how 'when playing gigs with people I know, we can often play complementary phrases and we direct what each other will play, it all happens in the milliseconds, just like day-to-day speech, and its beautiful, because it is unspoken. When I do session work, or if I'm called for a gig when I've never met the musicians, there is still an unspoken language (in jazz/funk at least) when we all understand what is being communicated'.

Elisabeth's childhood ambition was to be an orchestral musician, and unlike Sarah she immediately connected with the music she learned as a young child. She describes how she has always been able to experience and express emotions through classical music. Elisabeth's description of long childhood summers spent mastering complex piano compositions resonates with Csikszentmihalyi's description of autotelic individuals who achieve a rewarding balance between work and play. Elisabeth describes the 'massive' flow she has experienced playing in symphony orchestras, telling how 'with music as an artform—you are living in the past, the present and the future at the same time'. She believes that music cuts across boundaries between people, describing how 'you can have very intimate experiences with someone you don't know at all well when making music with them'. For Elisabeth music is a universal language and musicality is a core element of humanity that transcends neurological, cultural and political divides.

In the following chapter, motivation, experience and individual differences will be further explored in the context of Andrew, James, Stéphane and Will's accounts of their musical experiences and development. All of these musicians experience hyperacusis, or co-occuring hyperausis and synaesthesia and the ways these influence musical development and creativity will be explored.

References

1. Murray, D., Lesser, M., & Lawson, W. (2005). Attention, monotropism and the diagnostic criteria for autism. *Autism, 9*(2), 139–156. https://doi.org/10.1177/1362361305051398. PMID: 15857859.
2. Murray, F. (2018). Me and Monotropism: A unified theory of autism. *The Psychologist, 32*, 44–49.
3. Csikszentmihalyi, M. (1975). *Beyond boredom and anxiety.* Jossey-Bass. [Reprinted in 2000 with a new introduction].
4. Csikszentmihalyi, M., & Csikszentmihalyi, I. S. (Eds.). (1988). *Optimal experience: Psychological studies of flow in consciousness.* Cambridge University Press.
5. Milton, D. (2017). Going with the flow: autism and 'flow states'. In *Enhancing lives – reducing restrictive practices'*, 28 Sept 2017, Basingstoke, UK (KAR id:63699).
6. Rapaport, H., Clapham, H., Adams, J., Lawson, W., Porayska-Pomsta, K., & Pellicano, E. (2023). "In a state of flow": A qualitative examination of autistic Adults' phenomenological experiences of task immersion. *Autism in Adulthood, 6*, 362.
7. Csikszentmihalyi, M. (1990). *Flow: The psychology of optimal experience.* Harper & Row.
8. Tan, L., & Sin, H. X. (2019). Flow research in music contexts: A systematic literature review. *Musicae Scientiae, 25*(4). https://doi.org/10.1177/1029864919877564
9. Heasman, B., Williams, G., Charura, D., Hamilton, L. G., Milton, D., & Murray, F. (2024). Towards autistic flow theory: A non-pathologising conceptual approach. *Journal for the Theory of Social Behaviour.* (In Press).
10. American Psychiatric Association. (2013). *The diagnostic and statistical manual of mental disorders* (5th ed.; DSM–5). American Psychiatric Association.

11. Milton, D. E. M. (2012). On the ontological status of autism: the 'double empathy problem'. *Disability & Society, 27*(6), 883–887. https://doi.org/10.1080/09687599.2012.710008
12. Sasson, N. J., Faso, D. J., Nugent, J., Lovell, S., Kennedy, D. P., & Grossman, R. B. (2017). Neurotypical peers are less willing to interact with those with autism based on thin slice judgments. *Scientific Reports, 7*, 40700. https://doi.org/10.1038/srep40700. PMID: 28145411; PMCID: PMC5286449.
13. Sheppard, E., Pillai, D., Wong, G. T., Ropar, D., & Mitchell, P. (2016). How easy is it to read the minds of people with autism spectrum disorder? *Journal of Autism and Developmental Disorders, 46*(4), 1247–1254. https://doi.org/10.1007/s10803-015-2662-8. PMID: 26603886.egulated." (McDonnell and Milton, 2014).
14. Cross, I. (2014). Music and communication in music psychology. *Psychology of Music, 42*(6), 809–819. https://doi.org/10.1177/0305735614543968
15. Swanwick, K., & Tillman, J. (1986). The sequence of musical development: A study of children's composition. *British Journal of Music Education., 3*(3), 305–339. https://doi.org/10.1017/S0265051700000814

7

Musical Journeys 2: Heightened Perceptual Experience and Musical Creativity

Abstract In this chapter the effect of auditory hypersensitivity and synaesthesia on musical experience and development will be explored through first-person accounts of autistic musicians. These show that auditory and other hypersensitivities can create difficulties in live rehearsal and performance, but may also motivate creative exploration of musical form and enhance multi-sensory experiences of music.

Keywords Heightened perceptual experience and musical creativity • Auditory hypersensitivity • Synaesthesia • Sensory disturbance • Neurodiversity

Sensory Disturbance in Autism

Hyper- and/or hyporeactivity to sensory input were first included in diagnostic criteria for autism in DSM-5 (2013) [1] and research has documented difficulties in sensory, vestibular and proprioceptive modalities in autistic children and adults [2]. Sensory processing disturbances emerge early in development, and whilst patterns of disturbance and their levels of severity may change over time, they frequently persist into adulthood [3].

The precise prevalence of sensory disturbance in autism is difficult to determine given heterogeneity and constraints relating to methods of assessment. However, a recent analysis of data from a sample of around twenty-five thousand children reported that 74% of autistic individuals experienced significant sensory difficulties [4]. Hyperacusis is characterised by hypersensitivity or decreased tolerance to sounds and is believed to affect approximately 70% of autistic people [5]. One of the six autism+music experts (Sarah) has absolute pitch, a characteristic defined by an ability to identify isolated musical tones, without the need for a reference tone. In common with the majority of musicians, the remaining five experts report relative pitch, where tonal context is necessary for pitch identification.

These musicians all report exceptionally fine-grained pitch discrimination, a characteristic that has been identified in some experimental studies of autism and explained in the context of theoretical accounts of autistic strengths [6–8]. However, in the case of these particular individuals it is impossible to determine whether fine-grained pitch perception results from perceptual strengths characteristic in autism, musical training that results in changes in pitch processing [9] and patterns of connectivity in structural and functional brain networks [10], or both.

Three of the six experts have synaesthesia, a condition in which external stimulation in one modality (e.g. visual, auditory, olfactory) induces involuntary concurrent perceptual experiences in another modality. Research into synaesthesia suggests a prevalence rate of approximately 4% in the general population [11], and in one study of autistic adults, this figure rose to 18.9% [12]. Synaesthesia has been shown to be more common amongst musicians, writers and artists [13] and the list of composers with this condition includes Rimsky-Korsakov, Liszt, Sibelius, Scriabin, Beach, Ligeti and Messiaen. In one study Ward, Thompson-Lake, Ely and Kaminski (2008) [14] reported a link between an individual's musical engagement and the type of synaesthesia they experience. Within their sample of 82 participants with different forms of synaesthesia, those who reported visual experiences in response to music were more likely to play a musical instrument than those with other forms of synaesthesia.

Auditory Hypersensitivity and Musical Development

In the study by Korošec, Osika and Bojner-Horwitz (2022) [15] some autistic adults used self-selected music as a way of 'screening out' unpleasant and intrusive environmental stimuli. For example, one participant told how 'if I forget to bring my headphones when I am commuting, it is of course going to be a rough journey'. These difficulties present significant challenges for autistic musicians, who in many situations are unable to adopt such a strategy.

Sensory difficulties influence both day-to-day experiences of music and trajectories of musical development in autism. In his account of his musical experiences, Will has described the importance of playing with others, reporting that 'since playing with others, music has become an intense interest rather than just boxing myself into one genre'. However he also reports 'playing with others, I struggled with for a very long time as my ability to follow cues quickly while dealing with lights, people watching, and the complexity of the noise caused me huge burnout'. Will's early attempts to manage high levels of 'sensory noise' sometimes resulted in difficulties with other players, and he reports that 'in jam sessions I follow my pure instinct, however, sometimes that leads to not listening to other players due to dissociation. When I explained this to disgruntled players, they just accused me of being rude and egotistic. This put me off playing with others for a fair while'.

Andrew's early musical development was powerfully influenced by his auditory hypersensitivities. These included both hypersensitivity to sound volume, characteristic in hyperacusis, and exceptionally fine-grained discrimination of sounds. His earliest formal musical training was on the piano, and his initial experiences of learning this instrument left him feeling very confused. For example, he found that chords categorised as consonant within music theory sounded fundamentally dissonant when played on this instrument. He recalls being asked, during an aural training session, to identify the chord that comprises the first, third and fifth degrees of the scale (a tonic triad). This is a consonant chord, but for Andrew the sensory disturbance resulting from dissonance within the third and fifth pitch intervals in equal temperament tuning resulted in a sound that could not be categorised in this way.

Since the Romantic period in Western classical music, keyboard instruments have been uniformly tuned to the 'equal temperament scale', which divides the octave into 12 semitones of equal size. The advantage of 'equal temperament' over other tuning systems is that distances between adjacent tones are equivalent, both within and across keys, whilst a disadvantage is that pitch intervals, apart from octaves, consist of irrational-number frequency ratios, and are thus impure. In 'pure' or 'just' tuning systems, intervals and chords are comprised of tones from a single harmonic series that are positive integer multiples of a common fundamental frequency. This means that the frequencies of the tones are far closer to the harmonics than is the case in equal temperament scales.

As a young person Andrew did not encounter anyone else who experienced discomfort in response to equal temperament scales, but two early musical experiences enabled him to make sense of this. The first of these experiences occurred when he attended a performance in 1974 of Peter Maxwell Davies's composition 'Ave Maris Stella'. This composition is based on plainsong chant and scored for flute, clarinet, piano, viola, cello and marimba. The marimba is a larger xylophone, a percussion instrument consisting of wooden bars that are struck by mallets. The resonating tubes below the bars amplify particular harmonics of its sound. In one movement of 'Ave Maris Stella' there is a marimba solo with very low notes and Andrew described the joy of experiencing the overtone series as truly consonant for the very first time. The second formative experience occurred when Andrew's piano teacher introduced him to 'Mikrokosmos', a series of piano pieces written by the Hungarian composer Béla Bartók (1881–1945). Bartók's highly individual compositional style was somewhat influenced by classical and modernist composers such as Strauss, Debussy, Stravinsky and Schoenberg, which he combined with the folk music of Rumania, Hungary and Slovakia. This music is rich in chromaticism, dissonance, irregular rhythms and the use of intervals such as tritones. Andrew loved learning these pieces. His perception of the interval relations within equal temperament scales meant that he experienced the piano as a dissonant instrument and found that 'Mikrokosmos' was perfectly suited to it.

Like many young music students of his generation, the first piano pieces Andrew learned were written during the Classical and Romantic periods, and as a young musician he experienced tension over the question of whether he should write in the style of earlier, 'proper' composers or in a style that 'appealed to his ear'. However, music is a constantly evolving cultural form and Andrew's musical development occurred in a period of great change. The year 1952 had witnessed the premier of John Cage's 4'33", and this sparked considerable and ongoing debate about what music is and how we listen to it. 4'33" is scored for any instrument or instrumental grouping and is written in three movements. During the performance the musicians do not make any sounds on their instruments, and the 4 minutes 33 seconds framed what many audience members perceived as silence. However, as Cage later claimed: 'They missed the point. There's no such thing as silence. What they thought was silence, because they didn't know how to listen, was full of accidental sounds'. For Cage the composition represented a temporal window within which listeners could experience a unique soundscape that encompassed sounds made by the wind and rain and by the actions of the people within the concert hall. John Cage's work had a profound impact on the generations of composers that followed him, and for Andrew this unification of stimuli traditionally separated into 'musical' and 'environmental' sound categories had great personal resonance.

Synaesthesia, Sensory Disturbance and Evolving Musicality

Since the beginning of his career, Andrew's creative output has been influenced by his synaesthesia, and a recent example of this work will be described in the following section. Although it is clear that Andrew's sensory and perceptual traits have influenced his musical development, the MGIM model of talent [16] implicates multiple interacting factors in its emergence. Andrew's interests in the arts, philosophy and mathematics, and the musical culture in which he developed, have played an equally important role in shaping his approach to composition, improvisation and performance. His beliefs about what music and musicality are reflect these interacting influences. He writes, 'It is this combination of the objective understanding of music coupled with the innately subjective character in its creation that, for me, renders it both an art and a science at once'. 'It seems to me that music itself demands both artistic and scientific thinking in order to be complete. The knowledge it contains is therefore hybrid and thoroughly interdisciplinary. It is an understanding of the human self, but also of the nature of the world, an acceptance of infinite variation, but also the identification of eternal truths. In other words, it is both positivistic and phenomenological, in fact a perfect example of an interdisciplinary science' [17].

Creative Journeys

Born in 1957, Andrew's adolescence and early adulthood coincided with a period of unprecedented evolution in musical culture across Europe and North America. During the second half of the twentieth century, the output of composers working within established tonal, experimental and avant-garde schools of composition continued to influence new work. There was also an explosion of new musical movements that built on and/or radically diverged from these established traditions. During this period, 'classical' music was increasingly influenced by jazz, blues, African and Eastern music, the visual and theatre arts, and different schools of political and philosophical thought. In London during the 1970s and 1980s, there was a proliferation of small, composer-driven ensembles performing music that experimented with form, style, instrumentation and modes of presentation.

In 1981 Andrew founded the George W. Welch ensemble, and its development powerfully captures the spirit of the time. In an article documenting the inauguration and development of the ensemble [18] Andrew describes the communal and multidisciplinary nature of its conceptualisation and development. Seeded in early collaborations with fellow musicians and artists, new members of the ensemble brought their own interests and expertise in the fields of art, philosophy, popular music, jazz and film music. A crucial influence on the developing ethos of the ensemble was Marcel Duchamp (1887–1968), whose work motivated the visual art that became a crucial part of the ensemble's creative output. The George W. Welch ensemble regularly performed in prominent music venues in the UK until its final performance in the mid-1990s.

Auditory Perception Synaesthesia and Neurodiversity

In his early 50s, Andrew was diagnosed with Meniere's disease, a condition that is associated with recurring episodes of vertigo, hearing loss, tinnitus and diplacusis (in which the two ears hear different pitches). These difficulties made it very challenging to continue performing and Andrew began to focus on other areas of music. An important strand in his current creative and academic work is concerned with exploring differences in the ways people experience their perceptual worlds. In 2019 he launched the Aural Diversity project for and by people with a wide range of hearing differences.

During the Covid lockdowns in 2020 and 2021, Andrew was commissioned by the BBC, in partnership with Arts Council England, Arts Council of Northern Ireland, Arts Council of Wales and Creative Scotland, to compose a piece for inclusion in their 'Culture in Quarantine' series. This composition, 'Spectrum Sounds',[1] comprised seven compositions, all of which link to different points on the colour spectrum. Andrew's stated aim in composing this piece was to 'draw out the richness and beauty of sound colours in ways that may be appreciated by the target audiences of neurodivergent and aurally diverse people, and anybody else who has a willingness to listen in a different way'. The format of 'Spectrum Sounds' is multi-sensory with audio and visual versions that include images and words that create links with the audio files but can be experienced without the sounds. There is also an immersive version in

---

[1] Link to 'Spectrum Sounds': https://andrewhugill.com/spectrumsounds/.

which multiple devises can be connected using wireless technology and the compositions can be experienced spatially.

Each of the different pieces included in 'Spectrum Sounds' was written for specific musicians and had a unique working process. For example, 'Fluted Orange Turbulence' was written for the autistic flautist, composer/sonic artist and autism advocate Anya Ustaszewki. Andrew experiences the flute as a fundamentally colourless instrument and the synaesthesic link between 'orange' and 'turbulence' in this composition reflects Anya's personal responses to sounds. Although Anya has hyperacusis she enjoys turbulent sounds, for example in heavy metal and some rock music, so Andrew created an archive of electroacoustic sounds that Anya ranked along a 'love, like, hate' continuum. She then recorded her flute responses to the sounds she 'loved' and 'liked' and these were integrated with the electroacoustic sounds in the final composition. A second piece, 'Ice Hole Trumpet', was written for Robyn Steward, an autistic musician, author, researcher and autism advocate.

'Ice Hole Trumpet' is directly linked with Andrew's synaesthesia, in this case in the association between high trumpet playing on or around F#, wind sounds and strong blue colours. The auditory stimuli that Andrew selected for use in the composition were sounds of wind blowing through ice holes in a frozen arctic terrain, and these were minimally processed to blend with the sounds of the trumpet. Robyn is primarily an improvising musician and worked from a written description of the nature of the wind sounds and suggestions about how she should formulate her responses (e.g. timbre, pitch range, register) to them in the three sections of the piece. Robyn's recordings of her improvisations were then edited and integrated with the wind sounds and the visual representation of the landscape.

A third piece within the suite, Rook+Cello, was written for Elisabeth Wiklander (biographical details in Chap. 5). For Andrew the rook, the minor sixth interval on the cello, the note E and the Phrygian mode evoke vibrant purple colours and these associations are emphasised in this highly lyrical composition. Rook+Cello is written in the Phrygian mode and notated, and the musical score is integrated into the multi-sensory version of the composition.

James and Stéphane

Both James and Stéphane are at relatively early stages in their professional careers though both have been actively engaged with music for as long as they can remember. Looking back on their earliest musical experiences, James vividly recalls the joy and freedom of dancing to the music of Prokofiev and the contrasting experiences of the first 'music' they played on the recorder. Whilst this experience was less than exciting, the recorder lessons provided opportunities for 'musical' performance, which James loved, and paved the way for clarinet lessons and exposure to the rich and varied clarinet repertoire. James made rapid progress on this instrument and within a few years had learned works by Weber, Lutosławski, Poulenc, Brahms and Mozart.

During their adolescence James was fortunate in encountering several musicians who supported and inspired their learning and musical exploration. James's clarinet teacher was a virtuosic saxophonist and composer who encouraged them to explore different compositional styles and techniques on their instrument. She introduced them to jazz and this enabled them to explore another musical genre and develop their improvisational skills. Although James initially taught themselves the piano, they went on to study with a piano teacher who was also to become an important influence in their musical development. This teacher had completed a PhD in contemporary piano music, and under her guidance James's appreciation of the creative potential of this instrument was considerably enriched.

James found the period leading up to university entrance exams challenging as their school did not offer Music A-level and they had to study without academic support. However, James's mother arranged private composition lessons, and these were to have a lasting impact on their musical journey. James recounts how their composition teacher lent them a CD of the Kronos Quartet playing works by Steve Reich. This music was radically different to anything James had previously encountered. Minimal musical elements (e.g. short phrases) were constantly repeated and transformed in harmonic and rhythmic shifts. James's composition teacher also introduced them to her own experimental work, further enriching their experience of music that was both 'contemporary and outside of the academy'.

James recalls how, after reviewing one of their compositions, their teacher had asked, 'Where are you in all this?' This question was to have long-term resonance. James completed their undergraduate degree in a traditional academic environment and amongst other skills developed a strong grasp of compositional techniques. However, whilst they could now compose music in the styles of many traditional and contemporary composers, they simply didn't want to. At this point they had begun to understand how social pressures around 'normativity' had meant that they no longer experienced music in the 'creative, unrestrained and embodied' way that they had as a child. For James, the period of their PhD marked a search for musical authenticity.

James's interests are intense and intellectually and emotionally rewarding. As a young child they had possessed exceptional mathematical as well as musical skills, and during their adolescence they developed interests in science, literature and philosophy. These interests powerfully shaped their perception of the world and the things they want to express in their work. They are very interested in Daoist philosophy that conceptualises reality as a process in which things come together whilst constantly transforming. Like Andrew, James does not believe that music, mathematics, science, literature and philosophy should be viewed as separable domains of human knowledge and experience. They also reject mind/body dualism, arguing that 'What is important to me about expressing myself as an artist, as an Autistic, and as a synaesthete: that there is an embodied, sensory language that arrives out of our engagement with the world and situates us in the world'.

The first two compositions in their ongoing project 'The Games of Life', 'The Game of Life' along the River Ouse (slightly symmetrical) and 'The Game of Life' across Clifton Ings (a watery walk), communicate James's sense of the rich complexity of human experience. The 'Game of Life' is a widely used game that simulates the complex patterns that occur as a person progresses through the different stages of their life. Initially devised by John Conway, a British mathematician, in 1970, the life journey is determined by the individual's initial state and their interactions with eight 'neighbours', represented in adjacent cells within a two-dimensional orthogonal grid. Interactions between cells are guided by transitions that draw on rules from population dynamics and determine whether a cell lives or dies. These pieces are created in stages. They first recorded the walk on their mobile phone and then used the Game of Life to control which frequencies in the sound spectrum were prominent at different points in time. The listener/observer of the piece does not see the walker but shares their experiences as they navigate a path that runs adjacent to the river Ouse.

At points on the journey the changing views of the river, path, trees and grass, and sounds of footsteps, running water and birds, convey a sense of unity, constancy and the passing of time. However, at other points, this feeling is disturbed by moments of foreboding as ambient, rumbling natural and mechanical aspects of the soundscape become more prominent and take on a different character. In this composition the listener/observer shares the walker's lived experiences of sensations and events that are not random but are multiple and interacting. Whilst James's compositional style has always been eclectic and is likely to change and evolve over the coming years, their belief in the embodiment of human experience, and the importance of creatively engaging with complexity, is likely to continue to motivate their work.[2]

Of all four experts described in this chapter, Stéphane is at the earliest stage of her career. In Chap. 5 her first memories involving musical exploration were described, and it is important to contextualise these in her later subjective experiences of music. Writing on the way synaesthesia has shaped her perception of music she says, 'I believe my perception has been influenced in such a way that, to me, music is volumetric. Thanks to synaesthesia, music "appears" to me as a spatial and multi-layered scenography. Elements of the music (rhythm, melodic features, silences, the timbre of instruments and/or voices, vibrations, frequencies, etc.) express themselves through a wide range of shapes, colours and volumes, which altogether constitute a 3D space with foreground elements, background elements, floating and dynamic elements, textures, and kinesthetics (sense of speed, acceleration, etc.). Music is thus a very spatial experience'.

---

[2] James composes and writes under the pen name Morgaine Rire. The ongoing 'The Games of Life' project can be found at https://youtube.com/playlist?list=PLoq4pELIanXmP13o8cefgHfl7wVAyd tfp&si=xvIhHmzFqwwFPyFn, and a larger catalogue of compositions and writings can be found at jamesredelinghuys.com.

During the course of her childhood Stéphane mastered what was for her the daunting task of learning musical notation. Although the difficulties she experienced are likely to have been exacerbated by her dyslexia and dyscalculia (difficulties with numbers), a far greater problem was that musical notation simply did not map onto music as she perceived and experienced it. She writes, 'Describing music with mere black-and-white notes and its daunting array of musical grammatical signs is akin to describing the taste of a pear without the ability to experience the fruit with all the human senses. In my experience as an autistic individual with synaesthesia, conventional musical notation barely captures the experiential richness of musical composition'.

Like James, Stéphane experiences musical composition as a 'full mind-body system experience'. She describes how sounds are first imagined and organised in a way that enables them to 'talk' or 'clash' with each other. This is then followed by an assessment of the effects these different sound combinations and configurations have on the ear, and throughout the body. Stéphane's work is motivated by a desire to share her vision with others, and like Andrew before her, she aims to capture and share the full sensory and emotional spectrum of music from a synaesthetic perspective.

Describing her future goals, she describes how she aims to develop an immersive (virtual or mixed reality) digital tool that will enable synaesthetic experiences in individuals who do not automatically have them, and facilitate musical expressivity and creativity in individuals for whom musical notation represents a barrier to these processes. As a musician with training in psychology, Stéphane is also interested in brain plasticity, particularly in understanding how synaesthesia and music together influence brain development. Advances in our understanding of sensory systems and their interactions, as well as developments in the technological tools available to composers and sonic artists, offer great promise for the future. Considering these developments in the context of what has been learned from the first-person accounts of autistic composers, it becomes clear that neurodivergent artists will continue to play an important role in this future.

References

1. American Psychiatric Association (Ed.). (2013). *Diagnostic and statistical manual of mental disorders* (5th ed.). American Psychiatric Association.
2. Schaaf, R., & Lane, A. (2014). Toward a best-practice protocol for assessment of sensory features in ASD. *Journal of Autism and Developmental Disorders, 45*, 1380–1395. https://doi.org/10.1007/s10803-014-2299-z
3. Tavassoli, T., Miller, L. J., Schoen, S. A., Nielsen, D. M., & Baron-Cohen, S. (2014). Sensory over-responsivity in adults with autism spectrum conditions. *Autism, 18*(4), 428–432. https://doi.org/10.1177/1362361313477246. Epub 2013 Oct 1. PMID: 24085741.
4. Kirby, A. V., Bilder, D. A., Wiggins, L. D., Hughes, M. M., Davis, J., Hall-Lande, J. A., Lee, L.-C., McMahon, W. M., & Bakian, A. V. (2022). Sensory features in autism: Findings from a large population-based surveillance system. *Autism Research, 15*, 751.
5. Danesh, A. A., Lang, D., Kaf, W., Andreassen, W. D., Scott, J., & Eshraghi, A. A. (2015). Tinnitus and hyperacusis in autism spectrum disorders with emphasis on high functioning individuals diagnosed with Asperger's syndrome. *International Journal of Pediatric Otorhinolaryngology, 79*, 1683–1688.
6. Mottron, L., Bouvet, L., Bonnel, A., Samson, F., Burack, J. A., Dawson, M., & Heaton, P. (2013). Veridical mapping in the development of exceptional autistic abilities. *Neuroscience and Biobehavioral Reviews, 37*(2), 209–228. https://doi.org/10.1016/j.neubiorev.2012.11.016. Epub 2012 Dec 5. PMID: 23219745.
7. Baron-Cohen, S., & Lombardo, M. V. (2017). Autism and talent: the cognitive and neural basis of systemizing. *Dialogues in Clinical Neuroscience, 19*(4), 345–353. https://doi.org/10.31887/DCNS.2017.19.4/sbaroncohen. PMID: 29398930; PMCID: PMC5789212.
8. Happé, F., & Frith, U. (2006). The weak coherence account: Detail-focused cognitive style in autism spectrum disorders. *Journal of Autism and Developmental Disorders, 36*(1), 5–25. https://doi.org/10.1007/s10803-005-0039-0. PMID: 16450045.
9. Zioga, I., Di Bernardi, L. C., & Bhattacharya, J. (2016). Musical training shapes neural responses to melodic and prosodic expectation. *Brain Research, 1650*, 267–282. https://doi.org/10.1016/j.brainres.2016.09.015. Epub 2016 Sep 10. PMID: 27622645; PMCID: PMC5069926.

10. Leipold, S., Klein, C., & Jäncke, L. (2021). Musical expertise shapes functional and structural brain networks independent of absolute pitch ability. *Journal of Neuroscience, 41*(11), 2496–2511.
11. Simner, J., Mulvenna, C., Sagiv, N., Tsakanikos, E., Witherby, S. A., Fraser, C., Scott, K., & Ward, J. (2006). Synaesthesia: The prevalence of atypical cross-modal experiences. *Perception, 35*(8), 1024–1033. https://doi.org/10.1068/p5469. PMID: 17076063.
12. Baron-Cohen, S., Johnson, D., Asher, J., Wheelwright, S., Fisher, S. E., Gregersen, P. K., & Allison, C. (2013). Is synaesthesia more common in autism? *Molecular Autism, 4*(1), 40. https://doi.org/10.1186/2040-2392-4-40. PMID: 24252644; PMCID: PMC3834557.
13. Ramachandran, V. S., & Hubbard, E. M. (2001). Synaesthesia--a window into perception, thought and language. *Journal of Consciousness Studies, 8*(12), 3–34.
14. Ward, J., Thompson-Lake, D., Ely, R., & Kaminski, F. (2008). Synaesthesia, creativity and art: What is the link? *British Journal of Psychology, 99*(Pt 1), 127–141. https://doi.org/10.1348/000712607X204164. PMID: 17535472.
15. Korošec, K., Osika, W., & Bojner-Horwitz, E. (2024). "It is more Important than food sometimes"; Meanings and functions of music in the lives of autistic adults through a hermeneutic-phenomenological Lense. *Journal of Autism and Developmental Disorders, 54*(1), 366–378. https://doi.org/10.1007/s10803-022-05799-2. Epub 2022 Nov 2. PMID: 36323990; PMCID: PMC10791771.
16. Ullén, F., Mosing, M. A., & Hambrick, D. Z. (2017). The multifactorial gene-environment interaction model (MGIM) of expert performance. In *The science of expertise* (pp. 365–375). Routledge.
17. Hugill, A. (2022). The continuous in motion: Music and/as science. *Interdisciplinary Science Reviews, 47*(2), 118–128. https://doi.org/10.1080/03080188.2022.2035101
18. Hugill, A. (2024). Systems and special interests: The formation of 'George W. Welch' in the early 1980s. Proceedings of *More Sounds, More Personalities: British Postmiminalism 1979–97*, Goldsmiths College, September 18th 2024.

8

Conclusions: Retrospective and Prospective

Abstract In this chapter I provide a brief summary of the main conclusions drawn from the work presented in the previous chapters. I outline limitations in cognitive models of talent in autism and propose a more complete account of the multiple factors involved in musical talent in autism. This re-appraisal of theoretical models has been motivated and informed by the accounts of autistic musicians, and I propose that future work should build on the insights that these provide. Research into musical talent in autism enriches our understanding of musicality as a complex human trait and may also contribute to ongoing debates about differences in profiles of communication strengths and difficulties distinguishing autistic and neurotypical people.

Keywords Retrospective • Prospective • Musical improvisation • 'Impairments' in communication

My aims in writing this book were twofold. The first of these was to re-examine the relationship between autism and musical talent in the context of theoretical and empirical work exploring the nature, development and functions of musicality. The second aim was to consider this question in the context of theoretical models developed by scholars working within the neurodiversity paradigm and based on the lived experience of autistic musicians. Research into musicality has shown that this is a complex, multi-component trait that shows considerable variability across individuals. The importance of understanding individual differences in musicality is accentuated in the case of autism, where there is such a high degree of variability in perceptual, cognitive, emotional, behavioural and social functioning skills. Musicality reflects complex interactions between an individuals' characteristics and the different components of music. Cognitive models of autistic talents which fail to acknowledge this complexity implicitly pathologize autistic talents.

Qualitative research, based on first-person reports of autistic people (described in Chap. 4), has radically advanced our understanding of musicality and musical engagement in autism. Research described in Chap. 4, has shown how autistic adolescents and adults utilise music for identity, intellectual, emotional and creative self-development, social-emotional connectivity, and personal well-being. There has been an explosion of interest in the therapeutic uses of music, and new interventions are increasingly implementing the important work traditionally carried out by music therapists. Korošec and colleagues [1] have developed a new theoretical framework for exploring the positive and negative consequences of musical engagement in autism. Lord and colleagues [2] have stressed the importance of research focused on ways to promote positive quality of life outcomes for autistic people, and well conducted research in the field of music offers great promise in this respect. We now need to build on methodologies used in music and well-being research to probe the question of why musical interests are enhanced and result in exceptional musical skills in some autistic people.

In Chap. 6, monotropism theory [3] and research into flow [4] were discussed in the context of lived experience in autism. Monotropism is characterised by a propensity to focus intensely on a small range of interests, and flow experiences are linked with monotropic interests. Music elicits attention and research described in Chap. 6 has revealed powerful links between multiple musical activities and flow experiences in neurotypical populations. This then suggests that there may be something of a 'fit' between characteristics of the domain and characteristics that have emerge in these accounts of autism. If, for a given individual, music is a topic of interest and their musical exploration culminates in the identification of a musical form that aligns with their musical preferences, flow will be experienced and may motivate levels of engagement that facilitate proficiency within the domain. A major strength of the monotropism and flow theories is that they are concerned with interests, motivation and reward, and these factors are important in explaining exceptional talents in neurotypical individuals. Research into musical prodigies without autism has revealed a tendency for intense focus [5] and an increased likelihood of experiencing flow in response to musical activities [6]. However, considered in contexts of the Multi-Factorial Gene-Environment Interaction Model (MGIM) [7] and the first-person accounts of musical development described in Chaps. 6 and 7, it must be concluded that monotropism and flow theories cannot provide a complete explanation for exceptional musical talents in autism.

What other factors need to be included to account for exceptional musicality in autism? I suggest that two others must be considered. The first of these relates to the emergence of exceptional musicality in the general population. Whilst interests, motivation and reward are linked with the emergence of talent, additional mechanisms (e.g. personality) are included in the MGIM model. A second factor relates more directly to autism. Atypical responses to sensory information are included in diagnostic criteria for autism but show considerable variability across individuals. Evidence presented in Chap. 7 shows that sensory processing differences can powerfully influence trajectories of musical development in some autistic individuals. However, this effect was not seen in all individuals. For the musicians described in chapter 6, development appeared to have been most strongly motivated by their appreciation of music's emotional, social, intellectual and aesthetic qualities. Moreover, evidence from Chap. 7 showed how developmental trajectories in musicians with marked sensory sensitivities were influenced by multiple mechanisms. In these cases, hypersensitivity to sensory information influenced the way music was experienced and in conjunction with personal factors, motivated creative and intellectual exploration of musical form and meaning. Theoretical models of exceptional musicality in autism must be able to account for the complex and heterogeneous nature of autism and the multiple mechanisms implicated in the emergence of exceptional skills in this domain. First-person accounts of autistic musicians, and research into the everyday uses of music in autistic adolescents and adults, presented in Chap. 4, raise questions about the validity of past theoretical and empirical work in this area. Importantly this work illustrates the futility of studying autistic people's talents outside the context of their lived experiences.

In Chap. 4 I described how different strands of research have informed our understanding of musicality as a complex human trait. This is particularly true for research based on first-person accounts of autistic people's musical experiences. This work confirms and extends our understanding of music's functions in everyday life, and the emotional, intellectual and social rewards afforded by musical engagement. Theoretical models of evolution have proposed interesting accounts of how our early ancestors' vocalisations became specialised for speech and song [8, 9], and this has led to interesting discussions about the type of information that is, and can be, communicated via speech and via music [10, 11]. Autistic people, their friends and family members as well as music therapists and educators have long been aware that some autistic people communicate more effectively via music than via speech. In Chap. 6 the question of communication was directly explored in the context of musical improvisation and everyday conversation in an autistic musician. Whilst diagnostic criteria for autism encompass 'impairments' in communication, exploration of musical improvisation revealed exceptional communication strengths. Communication is a complex psychological construct, and further exploration of this finding may help enrich our understanding of this important facet of music, and contribute to a better understanding of the nature of communication differences distinguishing autistic and neurotypical people.

References

1. Korošec, K., Backman Bister, A. & Bojner Horwitz, E. (in press). "A space to be myself": Music and self-determination in the lives of autistic adults. Psychology of Music.

2. Lord C, Charman T, Havdahl A, Carbone P, Anagnostou E, Boyd B, Carr T, de Vries PJ, Dissanayake C, Divan G, Freitag CM, Gotelli MM, Kasari C, Knapp M, Mundy P, Plank A, Scahill L, Servili C, Shattuck P, Simonoff E, Singer AT, Slonims V, Wang PP, Ysrraelit MC, Jellett R, Pickles A, Cusack J, Howlin P, Szatmari P, Holbrook A, Toolan C, McCauley JB. The Lancet Commission on the future of care and clinical research in autism. Lancet. 2022 Jan 15;399(10321):271–334. https://doi.org/10.1016/S0140-6736(21)01541-5. Epub 2021 Dec 6. Erratum in: Lancet. 2022 Dec 3;400(10367):1926. PMID: 34883054.
3. Murray, Dinah & Lesser, Mike & Lawson, Wendy. (2005). Attention, monotropism and the diagnostic criteria for autism. *Autism: The international journal of research and practice, 9*, 139–56. https://doi.org/10.1177/1362361305051398
4. Rapaport, H., Clapham, H., Adams, J., Lawson, W., Porayska-Pomsta, K., & Pellicano, E. (2023). "In a State of Flow": A Qualitative Examination of Autistic Adults' Phenomenological Experiences of Task Immersion. *Autism in Adulthood.*
5. Winner, E. (2000). The origins and ends of giftedness. *American Psychologist, 55*(1), 159–169. https://doi.org/10.1037/0003-066X.55.1.159
6. Marion-St-Onge, C., Weiss, M. W., Sharda, M., & Peretz, I. (2020). What Makes Musical Prodigies? *Frontiers in Psychology, 11.*
7. Ullén, F., Hambrick. D. Z., & Mosing, M. A. (2016). Rethinking expertise: A multifactorial gene-environment interaction model of expert performance. Psychol Bull.
8. Steven B. (1999). "The "Musilanguage" Model of Music Evolution", The Origins of Music, Nils L. Wallin, Björn Merker, Steven Brown.
9. Mithen, S. (2006). *The singing Neanderthals: The origins of music, language, mind, and body.* Harvard University Press.
10. Cross, I. (2014). Music and communication in music psychology. *Psychology of Music, 42*(6), 809–819. https://doi.org/10.1177/0305735614543968
11. Patel, A. D. (2008). *Music, language, and the brain.* Oxford University Press.

Index[1]

A

Absolute pitch/relative pitch, 10, 10n1, 20, 96
Autism prevalence, 3, 43
Autistic interests and talents, vii, 8–10, 60, 66, 75–83, 87–89
Autobiographical accounts, 11, 87
Autotelic personality, 87

B

Biomedical model of autism, 47
Biomusicology framework, 19–21, 25, 27, 60

C

Childhood schizophrenia, 38, 39, 62
Child prodigies, 76
Cognitive models of autism, 6–11
Communication in music, 89
Composers
    Bartok, 98
    Cage, 99
    with synaesthesia, 80, 96
Consonance/dissonance, 79, 80, 98
Co-occurring conditions, 41
Creativity, 20, 24, 62, 65, 67, 76, 88, 90, 95–106

---

[1] Note: Page numbers followed by 'n' refer to notes.

D

Definitions of music, 4, 5
Developmental models of autism, 9, 41, 44–47, 97
Diagnostic and Statistical Manual (DSM, American Psychiatric Association), 3, 39, 40
Diagnostic criteria for autism, autistic spectrum disorder, 3, 4, 65, 83, 95, 112
Double empathy problem, 62

E

Equal temperament, 80, 98, 99
Ethological framework (Timbergen), 21
Evolutionary theories of music, 4

F

Flow experience in autism, 87
Flow experience in music, 76, 86–88
Functions of music, 4, 11, 20, 25, 28–30, 66, 68, 88, 109, 111

G

Gender and autism, 42
Gender diversity and autism, 42
Genetics and environmental factors associated with autism, 4, 41, 44, 46, 47

H

Heritability, 43

Heterogeneity in autism, 4, 96
Hyperacusis, 65, 79, 96, 97, 102
Hyperlexia, 10, 10n1

I

International Classification of Diseases (ICD, World Health Organisation), 40

J

Jazz improvisation, 76, 103

M

Masking/camouflaging, 48
Meniere's disease, 101
Model of talent and expertise, 5
Monotropism theory, 85, 88
Motivation and reward, 60, 82
Music and aesthetic experience, 64, 67
Music and emotions, vi, 11, 27n1, 60, 61, 64, 66–68, 75–77, 81, 91, 106, 111
and brain correlates, 22
Musical instruments
 marimba, 98
 piano, v, 76, 77, 79, 80, 82, 90, 91, 97–99, 103

N

Neurodiversity, 7, 42, 47, 48, 101–102

P

Prospective longitudinal study designs, 44

S

Self-advocacy, 42

Sensory disturbance, 41, 66, 78, 79, 95–100
Social model of autism, 47, 48
Stigma, 48
Synaesthesia, 10, 10n1, 80, 81, 96, 97, 99–102, 105, 106

SPRINGER NATURE

GPSR Compliance

The European Union's (EU) General Product Safety Regulation (GPSR) is a set of rules that requires consumer products to be safe and our obligations to ensure this.

If you have any concerns about our products, you can contact us on ProductSafety@springernature.com

In case Publisher is established outside the EU, the EU authorized representative is:

Springer Nature Customer Service Center GmbH
Europaplatz 3
69115 Heidelberg, Germany

The manufacturer's authorised representative in the EU is Springer Nature Customer Service Centre GmbH, Europaplatz 3, 69115 Heidelberg, Germany. If you have any concerns regarding our products, please contact ProductSafety@springernature.com

Printed and bound by CPI Group (UK) Ltd, Croydon, CR0 4YY

08/04/2026

02085784-0001